Design Concepts for Seismic-Resistant Buildings

Design Concepts for Seismic-Resistant Buildings:

Quantitative Shaking Evaluations

By

Buntara Sthenly Gan

Cambridge
Scholars
Publishing

Design Concepts for Seismic-Resistant Buildings:
Quantitative Shaking Evaluations

By Buntara Sthenly Gan

This book first published 2023

Cambridge Scholars Publishing

Lady Stephenson Library, Newcastle upon Tyne, NE6 2PA, UK

British Library Cataloguing in Publication Data
A catalogue record for this book is available from the British Library

ISBN (10): 1-5275-9146-8
ISBN (13): 978-1-5275-9146-2

To

Rie, Kenta, our parents, families, and friends.

Thank you for the love and encouragement over the years

TABLE OF CONTENTS

PREFACE

State-of-the-art seismic-resistant building design and technology have undergone tremendous developments. In modern seismic-resistant building design codes, a building structure subjected to a strong earthquake can experience considerably large deformations without collapsing. The design is permitted up to the limited strength before the building collapses. According to data on casualties compiled by JMA, Japan Meteorological Agency of Japan, the number of humans injured due to strong earthquakes is higher than the number of deaths/missing. Likewise, the number of buildings collapsed was less than the partially damaged ones. This fact implies that the design of buildings that are based on continuously revised seismic-resistant standards has good earthquake resistance. However, it also concludes that the human deaths/injury casualties were not the results of the buildings' collapse but mainly due to the strong vibrations (*SHAKING*) initiated by the earthquake. In most cases, the tumbling of non-structural things, such as ceilings, bookshelves, furniture, etc., fell onto human beings, causing casualties. This book brings an augmentation of quantitative shaking criteria to the state-of-the-art seismic design practice to implement the effect of shaking to prevent human casualties due to strong earthquakes.

I would like to express my deepest gratitude to Mr. Chiaki Kageyama, the founder of Living Environment Design Office Co., Ltd. The company mainly constructs small-scale steel pile foundations for the customers of the residential house with a direct foundation. The first moment Mr. Kageyama came to my laboratory was in the spring of 2015. I took the challenges inspired by Mr. Kageyama asking me to create a program that can evaluate the **quantity of shaking** of houses during strong earthquakes. The program also can predict how effectively the steel pile foundation can reduce the quantity of shaking compared to the direct foundation. The criteria use the JMA, Japan Meteorological Agency's Japan's Seismic Intensity Level

(SIL), to calculate the quantity of shaking. JMA's SIL in Japan is totally different from other countries. The JMA's SIL describes the degree of shakings that human feels during earthquakes. In contrast, Richter Magnitude perception is still being adopted to describe earthquakes in many countries, even until I finished writing this book.

The Richter Magnitude is an energy quantity released from the source of an earthquake deep beneath the ground surface. Thus, the complexity of sedimentary soil layers will significantly influence the waves' propagation to the ground surface. Therefore, the Richter Magnitude does not describe the intensity of shakings that human feels during earthquakes.

However, the JMA's SIL is only a prediction of shaking of the ground surface surrounding the source of the earthquake. I elaborate this feature to the building floors and apply Artificial Intelligence (Deep Neural Network) to skip the computation time to predict the shaking instantly. Because processing time reduction is highly important for early warning system development. And the pending research that needs further attempt is to measure human emotions against shakings by measuring the electrical wave distribution of the human brain. This research will furnish the reliability of JMA's SIL scale of intensities made in the past based on coarse human perceptions. This brilliant idea brought me to proceed with the research to finalize all the outcomes of the latest stage into a book.

This book is intended for architecture, civil and structural engineering, geological, and information science students beginning a series of courses in earthquake engineering and disaster mitigation. Practitioners in designer and consultant companies are starting to learn or adopt the shaking quantity for real-time monitoring and to extend their research in measuring the quantitative shaking of either the existing or new design buildings. The governmental agency involves disaster mitigation plans, communication service providers, mass media planners, and device manufacturers in commercial, research, and development purposes of earthquake early warning systems.

The writing of this book was an enjoyable experience for me. I received benefits, encouragement, and support from many colleagues and my students, who have taught me how to explain difficult concepts in modest expressions through simple explanations. While it is impossible to name all of them, without their help and support, it could not have been possible to finish writing this book.

My sincere thanks to my respected former Emeritus Professor Fumio Nishino at the University of Tokyo for his philosophies in teaching and research, which have been useful in my academic and professional lives. I greatly appreciate Professor Mitsuharu Kurata at Nihon University for constantly motivating me to find the real "truth" in science.

I am indebted to Prof. Han Ay Lie at Universitas Diponegoro for asking me to be her co-promotor in supervising our Ph.D. student Mr. Nanang Gunawan Wariyatno. Dr. Nanang has passed his final defense on the basic part of this book!! Congratulation! I hope Dr. Nanang will continue our research to contribute his achievements to benefit his own country, Indonesia.

I want to thank my former master's students at the beginning of this work, namely, Mr. Yasuhiro Anzai, Mr. Naomichi Kobayashi, and Mr. Arata Chiba. My 2nd-grade master's student, Mr. Rito Yamakawa, for helping me carry out enormous structural dynamic analyses using the API interface of STRAND7. We are creating a databank of JMA's SIL shaking scales using thousands of earthquake data for AI implementations.

I appreciate the help and patience of the Commissioning Editor: Ms. Helen Edwards, the Typesetting Manager: Ms. Amanda Millar, and the Designer: Ms. Sophie Edminson of Cambridge Scholars Publishing, during this book's preparation and publication. It took a long time to finish this book since the contract was made around the autumn of 2018. Although the flooding incident in 2019, continued by the Covid-19 pandemic, came across, it cannot be an excuse for me to extend the submission of this book. Under the authority of Ms. Helen Edwards helped me finish this book.

Most books are not free from errors, especially those with many mathematical equations and numbers. I wish to thank in advance those readers who are willing to draw attention to any typos and errors using the following e-mail address:

[gan.buntarasthenly@nihon-u.ac.jp].

Buntara S. Gan
Koriyama, December 2022
Fukushima, Japan

CHAPTER ONE

ALL YOU NEED TO KNOW ABOUT EARTHQUAKES

Big earthquakes in history

At the beginning of the nineteenth century [1], records of human disasters and damage due to big earthquakes were collected. The magnitude of earthquakes was estimated based on historical and old archival reports. Table 1.1 shows the list of known big earthquakes with an estimated Richter magnitude of 8.5 or higher in history since 1500, sorted in ascending order of time.

Table 1.1 List earthquakes with a magnitude higher than 8.5 from 1500-2012 [2].

Date	Location	Naming	Magnitude
Dec. 16, 1575	Valdivia, Chile (then part of the Spanish Empire)	1575 Valdivia earthquake	8.5
Nov. 24, 1604	Arica, Chile (then part of the Spanish Empire)	1604 Arica earthquake (undocumented)	8.5
May 13, 1647	Santiago, Chile (then part of the Spanish Empire)	1647 Santiago earthquake	8.5
Oct. 20, 1687	Lima, Peru (then part of the Spanish Empire)	1687 Peru earthquake	8.5
Jan. 26, 1700	Pacific Ocean, USA, and Canada (then claimed by the Spanish Empire and the British Empire)	1700 Cascadia earthquake	8.7–9.2
Oct. 28, 1707	The Pacific Ocean, Shikoku region, Japan	1707 Hōei earthquake	8.7–9.3

Date	Location	Naming	Magnitude
Jul. 8, 1730	Valparaíso, Chile (then part of the Spanish Empire)	1730 Valparaíso earthquake	8.7
Oct. 17, 1737	Kamchatka, Russia	1737 Kamchatka earthquakes	8.5
Oct. 28, 1746	Lima, Peru (then part of the Spanish Empire)	1746 Lima–Callao earthquake	8.6
May 24, 1751	Concepción, Chile (then part of the Spanish Empire)	1751 Concepción earthquake	8.5
Nov. 1, 1755	Atlantic Ocean, Lisbon, Portugal	1755 Lisbon earthquake	8.5–9.0
Apr. 2, 1762	Chittagong, Bangladesh (then Kingdom of Mrauk U)	1762 Arakan earthquake	8.8
Mar. 28, 1787	Oaxaca, Mexico (then part of the Spanish Empire)	1787 Mexico earthquake	8.6
Nov. 19, 1822	Valparaíso, Chile	1822 Valparaíso earthquake (undocumented)	8.5
Nov. 25, 1833	Sumatra, Indonesia (then part of the Dutch East Indies)	1833 Sumatra earthquake	8.8
Feb. 20, 1835	Concepción, Chile	1835 Concepción earthquake	8.5
Feb. 16, 1861	Sumatra, Indonesia	1861 Sumatra earthquake	8.5
Aug. 13, 1868	Arica, Chile (then Peru)	1868 Arica earthquake	8.5–9.0
May 9, 1877	Iquique, Chile (then Peru)	1877 Iquique earthquake	8.5
Jun. 15, 1896	The Pacific Ocean, Tōhoku region, Japan	1896 Sanriku earthquake	8.5
Jan. 31, 1906	Ecuador – Colombia	1906 Ecuador–Colombia earthquake	8.8
Nov. 10, 1922	Atacama Region, Chile Catamarca Province, Argentina	1922 Vallenar earthquake	8.5
Feb. 1, 1938	The Banda Sea, Indonesia (then part of the Dutch East Indies)	1938 Banda Sea earthquake	8.5

Date	Location	Naming	Magnitude
Aug. 15, 1950	Assam, India – Tibet, China	1950 Assam–Tibet earthquake	8.7
Nov. 4, 1952	Kamchatka, Russian SFSR, Soviet Union	1952 Kamchatka earthquakes	9.0
Mar. 9, 1957	Andreanof Islands, Alaska, United States	1957 Andreanof Islands earthquake	8.6
May 22, 1960	Valdivia, Chile	1960 Valdivia earthquake	9.4–9.6
Oct. 13, 1963	Kuril Islands, Russia (USSR)	1963 Kuril Islands earthquake	8.5
Mar. 27, 1964	Prince William Sound, Alaska, United States	1964 Alaska earthquake	9.2
Feb. 4, 1965	Rat Islands, Alaska, United States	1965 Rat Islands earthquake	8.7
Dec. 26, 2004	Indian Ocean, Sumatra, Indonesia	2004 Indian Ocean earthquake	9.1–9.3
Mar. 28, 2005	Sumatra, Indonesia	2005 Nias–Simeulue earthquake	8.6
Feb. 27, 2010	Offshore Maule, Chile	2010 Chile earthquake	8.8
Mar. 11, 2011	The Pacific Ocean, Tōhoku region, Japan	2011 Tōhoku earthquake	9.1
Apr. 11, 2012	Indian Ocean, Sumatra, Indonesia	2012 Aceh earthquake	8.6

Figure 1.1 shows countries struck by strong earthquakes with magnitudes higher than 8.5 since the sixteenth century. Although Japan is well-known as the "Country of Earthquakes," it has only been hit three times. However, Chile has been struck by strong earthquakes twelve times.

Figure 1.1 Distribution map of earthquakes with a magnitude higher than 8.5 in history from 1500 to 2012.

Chile

The first documented Valdivia earthquake in 1575 led to the flooding of Valdivia city due to the excessive overflowing of rivers. In 1604, the next earthquake, which was not recorded, occurred in Arica. In 1647, another big earthquake struck Chile almost a century later. This earthquake in Santiago brought nearly all the buildings in the city to the ground. It is assumed that this earthquake was the most damaging in the history of Santiago city. Almost another century later, in 1730, the Valparaíso earthquake hit Chile again, which caused severe damage from La Serena to Chillán. The Valparaíso earthquake triggered a major tsunami along the coastline for more than 1,000 km. Just two decades after the Valparaíso earthquake, in 1751, the strongest and most destructive earthquake, Concepción, struck the central valley of the country and destroyed the cities of Concepción, Chillán, Cauquenes, Curicó, and Talca. Eighty years later, in 1835 (Figure 1.2), near the neighboring cities of Concepción and Talcahuano, the Concepción earthquake struck the area causing human disaster and triggering a tsunami in the coastal region.

Figure 1.2 The remains of the Cathedral in Concepción (a drawing by John Clements Wickham 1798–1864 Engraving: S. Bull fl. 1838–1846, from Wikimedia Commons).

In 1868, an earthquake struck a wide area around Arica, Tacna, Moquegua, Mollendo, Ilo, Iquique, Torata, and Arequipa, including the southern part of Peru [3] (Figure 1.3). The earthquake caused an estimated 25,000 human casualties and brought a tsunami to the coasts of Peru and Chile. In 1877, in the vicinity of the coastal area of the Tarapacá and Antofagasta regions, an overwhelming earthquake with a maximum intensity of XI (extreme) on the Mercalli intensity scale was reported. The earthquake's aftermath triggered a devastating tsunami on Fiji island, which caused 2,385 human casualties. Since the start of the twentieth century, the frequency of big earthquakes in Chile has significantly reduced. The 1922 Vallenar earthquake was followed by a destructive tsunami in the Atacama region, near the border with Argentina.

Figure 1.3 “The Late Earthquake in South America: Ruins of Arica, in Peru & Ruins of Arica from the Sea” October 24, 1868 (illustrated by London News South America, from Wikimedia Commons).

The earthquake caused extensive damage in a zone ranging from Copiapó to Coquimbo. Newspapers estimated more than 1,000 human casualties due to the earthquake, at least 500 of which were in Vallenar. Following the earthquake, a destructive tsunami killed several hundred people in coastal cities, especially in Coquimbo. Forty years later, the most powerful earthquake ever recorded in Chile was reported in Valdivia with a magnitude of 9.4–9.6, lasting less than ten minutes. This earthquake caused catastrophic damage due to its proximity to the hypocenter of the 1960 Valdivia earthquake (Figure 1.4).

The 2010 Chile earthquake occurred off the coast of central Chile, having a magnitude of 8.8, with intense shaking lasting for about three minutes. According to the United States Geological Survey (USGS), the cities experiencing the strongest shaking of VIII (Severe) on the Modified Mercalli intensity scale (MM) were: Concepción, Arauco, and Coronel. Chile's Seismological Service Concepción reported that the Concepción area experienced the strongest shaking at IX (Violent) on the MM scale. The earthquake generated a tsunami that devastated several coastal towns in south-central Chile and damaged the port at Talcahuano (Figure 1.5).

Figure 1.4 Earthquake damage to good quality, wooden-framed houses in Valdivia, Chile, 1960 (a photo credited by Pierre St. Amand, from Wikimedia Commons).

Figure 1.5 After the tsunami in Talcahuano, Biobío Region (a photo credited by Zelocaro from the source: Marcelo Caro Frías, in Wikimedia Commons).

South America

In Peru, the 1687 and 1746 earthquakes caused severe damage to the cities of Lima, Callao, and Ica. Similarly, the earthquakes brought about the subsequent tsunamis that caused human casualties. The 1787 Mexico earthquake also brought about a big tsunami that affected the coast at the boundary of Mexico. In 1906, the Colombia earthquake caused a destructive tsunami that affected the coast of Colombia with at least 500 human casualties. The 1922 Vallenar earthquake occurred in the Atacama region of Chile near the border with Argentina (Figure 1.6). This earthquake triggered a big tsunami that caused major damage to the coast of Chile.

Figure 1.6 Destruction in the Atacama Region in the 1922 Vallenar earthquake (a photo credited by Gustavo Bruzzone Rocco, in Wikimedia Commons).

Indonesia

The 1833 Sumatra earthquake occurred with an estimated magnitude of 8.8–9.2. The earthquake caused a large tsunami that flooded the island's southwestern coast. The 1861 Sumatra earthquake was the last in a sequence of earthquakes that ruptured adjacent parts of the Sumatran segment of the Sunda megathrust. It caused a shocking tsunami that led to several thousand human casualties. The 1938 Banda Sea earthquake happened with an estimated magnitude of 8.4 and a Rossi–Forel intensity of VII (very strong tremor). This oblique-slip event generated destructive tsunamis of up to 1.5 meters in height in the Banda Sea region, but no human casualties were reported. The 2004 Indian Ocean earthquake and tsunami occurred with the hypocenter off the west coast of northern Sumatra. It was an undersea megathrust earthquake recorded with a magnitude of 9.1–9.3, reaching a Mercalli intensity scale of up to IX in particular areas. The earthquake was caused by a rupture along the fault between the Burma Plate and the Indian Plate. The underwater seismic activity formed a series of tsunami waves of up to 30 meters (100 ft) in height. Communities along the surrounding coasts of the Indian Ocean were seriously affected, and the tsunamis caused an estimated 228,000 human casualties in fourteen countries (Figure 1.7). The Indonesian city of Banda Aceh was reported to have had the largest human casualties. The earthquake was one of the deadliest natural disasters in earthquake history. The earthquake was the third-largest ever recorded and had the longest duration of faulting, between eight to ten minutes. It caused the earth to vibrate as much as 10 millimeters (0.4 inches) and remotely triggered earthquakes as far away as Alaska. Its hypocenter was between Simeulue Island and mainland Sumatra.

Figure 1.7 A village near the coast of Sumatra lays in ruins after the tsunami struck (a photo credited by Photographer's Mate 2nd Class Philip A. McDaniel of the U.S. Navy, in Wikimedia Commons).

The 2005 Nias–Simeulue earthquake occurred off the west coast of northern Sumatra, Indonesia. More than 900 human casualties were reported, mainly on the island of Nias. The earthquake caused panic on the island, which had already been shocked by the massive tsunami triggered by the 2004 Indian Ocean earthquake. Still, this earthquake generated a relatively small tsunami that caused little damage. It was the third most powerful earthquake since 1965 in Indonesia.

Japan

The 1707 Hōei earthquake struck south-central Japan. The earthquake was the largest in Japanese history until the 2011 Tōhoku earthquake renewed the record after three centuries. The earthquake triggered a destructive tsunami that caused more than 5,000 human casualties in southwestern

Honshu, Shikoku, and south-eastern Kyūshū. The earthquake is also said to have triggered the last eruption of Mount Fuji 49 days later. About a century before the 2011 Tōhoku earthquake occurred, the 1896 Sanriku earthquake struck Japan in almost the same locations. The 8.5 magnitude earthquake brought about two tsunamis that destroyed about 9,000 residential houses and caused at least 22,000 human casualties (Figures 1.8, 1.9). At that time, the waves reached a record height of 38.2 m, until the record was renewed by more than 2m additional height in the 2011 Tōhoku earthquake.

Figure 1.8 Destruction in Kamaishi city, Iwate prefecture due to the Sanriku earthquake (a photo credited by Matsuchi Nakajima, 1850–1915, in Wikimedia Commons).

Figure 1.9 Destruction by the tsunami following the Sanriku earthquake (a photo credited by Imperial Household Ministry (present-day "Imperial Household Agency"), in Wikimedia Commons).

The 2011 earthquake off the Pacific coast of Tōhoku had a magnitude of 9.0–9.1, with the hypocenter approximately 70 kilometers east of the Oshika peninsula. The earthquake was the strongest earthquake ever recorded in Japan (Figure 1.10). The earthquake also triggered the most powerful tsunami waves that reached 40.5 m in height. The tsunami caused more than 15,000 human casualties and destroyed the Fukushima Daiichi Nuclear Power Plant nuclear facilities (Figure 1.11).

Figure 1.10 The Rikuzentakata, Iwate, after being swept by a tsunami on the Pacific Coast following the Tōhoku earthquake (a photo credited by Mitsukuni Sato, in Wikimedia Commons).

Figure 1.11 The Fukushima First Nuclear Power plants after the 2011 Pacific Coast Tōhoku earthquake and tsunami (a photo provided by Digital Globe, in Wikimedia Commons).

United States

The 1700 Cascadia earthquake occurred with an estimated Magnitude of 8.7–9.2 from Vancouver Island in Canada along the Pacific Northwest coast to the south to northern California. Two centuries later, no big earthquake with a Magnitude over 8.5 occurred until the 1964 Alaska, and 1965 Rat Islands earthquakes consecutively struck the United States. The 1964 Alaska earthquake, also known as the Great Alaskan earthquake, occurred in the south-central area of Alaska, destroying buildings and triggering tsunamis that caused human casualties (Figure 1.12).

Figure 1.12 Damage to Fourth Avenue in Anchorage, Alaska, caused by the 1964 Alaska earthquake (a photo credited by the U.S. Army, in Wikimedia Commons).

In the following year, the 1965 Rat Islands earthquake occurred within the Alaska region with a magnitude of 8.7. It triggered a 10m height tsunami on Shemya Island. However, the tsunami caused almost no destruction.

Russia

Three earthquakes that occurred off the coast of the Kamchatka Peninsula in the far east of Russia and the Soviet Union in 1737, 1952, and 1963 were big for the country and caused tsunamis. Little destruction and few human casualties were reported due to earthquakes and tsunamis.

Portugal

The first big earthquake with a magnitude of 8.5–9.0 in a European country was recorded in Portugal in 1755. The earthquake caused fires and tsunamis and almost destroyed the city of Lisbon (Figure 1.13). The number of human casualties was estimated to be between 10,000 to 100,000, making the earthquake one of the deadliest earthquakes in history.

Figure 1.13 Lisbon, Portugal, during the 1755 great earthquake. This copper engraving shows the city in ruins as tsunamis rush upon the shore (a photo retrieved from the Earthquake Engineering Online Archive, Jan Kozak Collection: KZ128, in Wikimedia Commons).

East Asia

The 1762 Arakan earthquake occurred at an estimated magnitude of 8.8 and the maximum estimated intensity of XI (Extreme) on the Mercalli intensity scale. The earthquake triggered a local tsunami in the Bay of Bengal and caused at least 200 human casualties. About two hundred years later, the 1950 Assam–Tibet earthquake occurred with a magnitude of 8.6. The earthquake was destructive in the Assam (India) and Tibet (China) regions, and approximately 4,800 human casualties were reported. The earthquake was notable as being the largest recorded earthquake caused by continental collision rather than subduction and was also notable for the loud noises produced by the earthquake and reported throughout the regions of both countries.

Building damage due to earthquakes

Physical damage due to earthquakes can be divided into two major categories:

1. Damage to the infrastructure of human-made structures
2. Damage to the geological elements supporting the structures below the soil and ground level

In the former, the physical damage to human-made structures is usually a result of design strength insufficiencies against the bigger earthquakes in history. In the latter, the physical damage to the supporting soil and ground is due to larger historical earthquakes and natural disasters in the area of concern.

Table 1.2 lists 25 earthquakes that caused approximate damage worth more than 500 million USD from 1976 to 2017.

Table 1.2 A list of earthquakes that caused damage worth 500 million USD from 1976 to 2017. [2].

Date	Location	Magnitude	Intensity	Damage (USD Mill)
Jan. 16, 1995	Japan: Kobe, Awaji-shima, Nishinomiya	6.9	11	100,000
May 12, 2008	China: Sichuan Province	7.9	9	86,000
Jan. 17, 1994	US: California, Northridge	6.7	9	40,000
Feb. 27, 2010	Chile: Maule, Concepción, Talcahuano	8.8	9	30,000
Oct. 23, 2004	Japan: Niigata Prefecture	6.6		28,000
Nov. 23, 1980	Italy: Avellino, Potenza, Caserta, Naples	6.9	10	20,000
Aug. 17, 1999	Turkey: Istanbul, Kocaeli, Sakarya	7.6	10	20,000
Apr. 15, 2016	Japan: Kumamoto, Oita	7	8	20,000
Dec. 7, 1988	Armenia: Leninakan, Spitak, Kirovakan	6.8	10	16,200
May 29, 2012	Italy: Emilia Romagna, Medolla, Mirandola, Cavezzo	5.9		15,800
Feb. 21, 2011	New Zealand: Christchurch, Lyttelton	6.1		15,000
Sep. 20, 1999	Taiwan: Nantou, Taichung, Taizhong	7.7	10	14,000
Jul. 16, 2007	Japan: West Coast	6.6		12,500
Dec. 26, 2004	Indonesia: Sumatra, Aceh, off the West Coast	9.1		10,000
Apr. 25, 2015	Nepal: Kathmandu; India; China; Bangladesh	7.8	8	10,000
Jan. 12, 2010	Haiti: Port-Au-Prince	7		8,000
Jun. 20, 1990	Iran: Rasht, Qazvin, Zanjan, Rudbar, Manjil	7.3	7	7,200
Oct. 8, 2005	Pakistan: Muzaffarabad, Uri, Anantnag, Baramula	7.6	8	6,680
Sep. 3, 2010	New Zealand: Christchurch	7	9	6,500
Sep. 19, 2017	Mexico: Mexico City, Morelos, Puebla	7.1		6,000

Date	Location	Magnitude	Intensity	Damage (USD Mill)
Jul. 27, 1976	China: Tangshan	7.5	11	5,600
Oct. 18, 1989	US: California: Loma Prieta	6.9	9	5,600
Oct. 10, 1980	Algeria: Northern	7.7	10	5,200
May 21, 2003	Algeria: Algiers, Boumerdes, Reghia, Thenia	6.8	10	5,000
Aug. 24, 2016	Italy: Accumoli, Arquata, Amatrice	6.2	9	5,000

From Table 1.2, the Great Hanshin-Awaji Earthquake that occurred on 17th January 1995 in the southern part of Hyogo Prefecture, Japan, tops the list. The earthquake had the largest intensity of 7 on the Japan Meteorological Agency (JMA) Seismic Intensity Scale. The earthquake shook the surrounding area for approximately 20 seconds. The earthquake's hypocenter measured about 17 km underneath the ground on the northern end of Awaji Island, about 20 km from the center of Kobe city. More than 6,000 human casualties were recorded, wherein 70% lived in Kobe city, which had a population of 1.5 million (Figure 1.14).

Figure 1.14 Damage to buildings in Kobe city due to the Hanshin-Awaji Earthquake (a photo credited by Akiyoshi Matsuoka, 1995, in Wikimedia Commons).

The damage was widespread and severe because the hypocenter was close to the city. Unrepairable structures damaged by the earthquake reached approximately 400,000 buildings, roads, bridges, and quays in the port of Kobe city. The earthquake also triggered about 300 fires over a large area of Kobe city. Human casualties of over 4,000 were reported in the cities and suburban areas of Hyogo Prefecture. Table 1.3 shows the statistical data of damage to more than 50,000 houses that occurred during earthquakes from 1976 to 2017.

Table 1.3 List of earthquakes that caused damage to 50,000 houses in 1976–2017 [2].

Date	Location	Magnitude	Intensity	Damage
May 12, 2008	China: Sichuan Province	7.9	9	5,360,000
Feb. 3, 1996	China: Yunnan Province: Lijiang	6.6	9	959,000
Nov. 6, 1988	China: Yunnan Province	7.3	10	412,000
May 25, 2008	China: Sichuan Province	6		400,000
Aug. 30, 2008	China: Sichuan Province	5.9		392,000
Jan. 26, 2001	India: Gujarat: Bhuj, Ahmadabad, Rajokot; Pakistan	7.7	10	339,000
Apr. 25, 2015	Nepal: Kathmandu; India; China; Bangladesh	7.8	8	299,588
Jul. 21, 2003	China: Yunnan Province	5.9		264,878
Jan. 16, 1995	Japan: Honshu: Kobe, Awaji-Shima, Nishinomiya	6.9	11	200,000
Jun. 12, 1730	China: Hebei Province	7.5	10	200,000
Nov. 26, 2005	China: Jiangxi: Jiujiang, Ruichang	5.2		150,000
Oct. 27, 1791	Japan: Mino-Owari	8.4		142,177
May 26, 2006	Indonesia: Java: Bantul, Yogyakarta	6.3		127,000
Jul. 9, 1979	China: Jiangsu Province: Liyang	5.4	8	113,909
Jan. 13, 2001	El Salvador	7.7	8	108,226

Date	Location	Magnitude	Intensity	Damage
Feb. 21, 2011	New Zealand: Christchurch, Lyttelton	6.1		100,000
Jul. 11, 1995	China: Yunnan Province: Lancing, Menglian; Thailand	6.8		100,000
Sep. 17, 1203	China: Shanxi Province	8	11	100,000
Jan. 12, 2010	Haiti: Port-Au-Prince	7		97,294
Sep. 6, 1733	China: Yunnan Province	8	11	87,620
Oct. 20, 1990	China: Gansu Province: Tianzhu	5.8	4	87,000
Apr. 15, 2016	Japan: Kumamoto, Oita	7	8	80,000
Sep. 26, 1997	Italy: Central: Marche, Umbria	6	10	80,000
Sep. 18, 1984	Turkey: E, Erzurum, Olur-Senkaya	6.4	8	75,000
Oct. 18, 1984	Turkey: E, Senkaya	5.3		75,000
Apr. 14, 1928	Bulgaria: Chirpan	6.8	10	74,750
Sep. 2, 1665	China: Gansu Province: Wushan, Gangu	6.5	9	74,748
Feb. 24, 2003	China: S. Xinjiang: Bachu	6.3		71,000
Jan. 10, 1998	China: Northeastern: Shangyi-Zhangbei, Hebei	5.7	8	70,000
Apr. 26, 1990	China: Qinghai Province: Ganghe-Xinghai	6.9		66,800
Jun. 28, 1948	Japan: Fukui	7.3		63,000
Feb. 23, 2001	China: Sichuan Province: Kangding, Yajiang	5.6		60,000
Oct. 28, 1607	Japan: Nankaido	8.4		77,000
May 22, 1960	Chile: Puerto Montt, Valdivia	9.5	12	58,622
Sep. 12, 2007	Indonesia: Sumatra	8.4		56,425
Nov. 3, 2006	China: Nei Mongol	4.7		55,886
Mar. 27, 1967	China: Hebei Province: Hejian, Dacheng	6.3	7	53,886
Sep. 7, 1999	Greece: Athens	6	9	53,000
Aug. 15, 2007	Peru: Ica, Pisco, Lima	8	8	52,154

Table 1.3 shows three consecutive earthquakes in {Sichuan Province, China, in 2008. The 2008 Sichuan Earthquake destroyed a total of more than 6 million houses. The earthquake's hypocenter was about 19 km deep and located about 80 km west-northwest of Chengdu, the provincial capital of Sichuan. The earthquake caused a fault of over 240 km, cracking the ground surface several meters across. The earthquake caused over 69,000 people to lose their lives and left about 4.8 million without their houses (Figure 1.15).

Figure 1.15 A collapsed residential building in Wenchuan in the aftermath of the Sichuan earthquake (a photo taken by Seoul Fire and Disaster Headquarters, 2008, in Wikimedia Commons).

Human casualties caused by earthquakes

Historically, earthquakes have shown the shortcomings of synchronous design methodologies and construction practices, resulting in structural failures and loss of human life. Post-earthquake investigations have improved engineering analysis, design, and construction practices. A summary of the historical development of earthquake engineering practice, showing how earthquake engineers have learned from past experiences, is shown in Table 1.4. Table 1.4 shows the statistical data of damage to more than 50,000 houses during earthquakes from 1976 to 2017.

Table 1.4 Earthquakes that caused human casualties in damaged houses from 1976 to 2017 [2].

Date	Location	Magnitude	Intensity	Damage
Dec. 26, 2004	Indonesia, Sumatra, Aceh, West Coast	9.1		227,899
Dec. 16, 1920	China: Gansu Province, Shanxi Province	8.3	12	200,000
Dec. 22, 856	Iran: Damghan, Qumis	7.9	10	200,000
Dec. 24, 893	Armenia: Devin; Iran: Ardabil		10	150,000
Sep. 1, 1923	Japan: Tokyo, Yokohama	7.9		142,807
Nov. 29, 533	Syria: Aleppo			130,000
Oct. 5, 1948	Turkmenistan: Ashkhabad	7.3	10	110,000
Sep. 27, 1290	China: Nei Mongol: Ningcheng	6.8	9	100,000
May 12, 2008	China: Sichuan Province	7.9	9	87,652
Aug. 31, 458	Turkey: Antakya (Antioch)		9	80,000
Oct. 31, 1667	Azerbaijan: Shemakha (Samaxi)	6.9	10	80,000
Dec. 28, 1908	Italy: Messina, Sicily, Calabria	7	11	80,000
Feb. 13, 1157	Syria: Halab (Aleppo)			80,000
Nov. 18, 1727	Iran: Tabriz		8	77,000
Oct. 8, 2005	Pakistan: Muzaffarabad, Uri, Anantnag, Baramula	7.6	8	76,213
Jun. 19, 1718	China: Gansu Province	7.5	10	73,000
Nov. 30, 847	Iraq: Al-Mawsil		10	70,000
Dec. 5, 1033	Israel: Ramala, Gaza, Nablus; Egypt		10	70,000
Aug. 16, 1868	Ecuador: Guayaquil, Ibarra⊠ Colombia: San Pablo	7.7	10	70,000
May 31, 1970	Peru: Northern, Pisco, Chiclayo	7.9	10	66,794

Date	Location	Magnitude	Intensity	Damage
Jan. 9, 1693	Italy: Sicily, Calabria, Catania	6	11	60,000
May 30, 1935	Pakistan: Quetta	7.5	10	60,000
Sep. 30, 587	turkey: Antakya (Antioch)		9	60,000
Nov. 30, 1101	Iran: Khorasan	6.5	10	60,000
Jan. 23, 1909	Iran: Silakor	7.7		55,004
Nov. 4, 1042	Iran: Tabriz	7.6	9	50,000
Jan. 3, 1739	China: Ningxia	8	10	50,000
Nov. 1, 1755	Portugal: Lisbon	8.5	11	50,000
Jan. 8, 1780	Iran, Tabriz	7.7	9	50,000
Dec. 26, 2004	Indonesia, Sumatra, Aceh, West Coast	9.1		227,899

The 2004 Indian Ocean earthquake struck Banda Aceh on the west coast of northern Sumatra Island, Indonesia. It caused several tsunami waves of up to 30m in height. The earthquake had a hypocenter off the west coast of northern Sumatra with a Richter magnitude of 9.1–9.3. The earthquake was caused by a rupture along the fault between the Burma Plate and the Indian Plate.

Figure 1.16 Tsunami inundation of the height of a house in Banda Aceh after the earthquake (a photo credited by Michael L. Bak, 2005, in Wikimedia Commons).

Seismic areas in the world

The *seismicity* map shows the locations of the *hypocenters* of earthquakes. The map usually shows all the earthquakes with a Magnitude larger than a specified value that occurred in a certain period. Figure 1.17 is a map showing the hypocenters of the 358,214 earthquakes from 1963 to 1998.

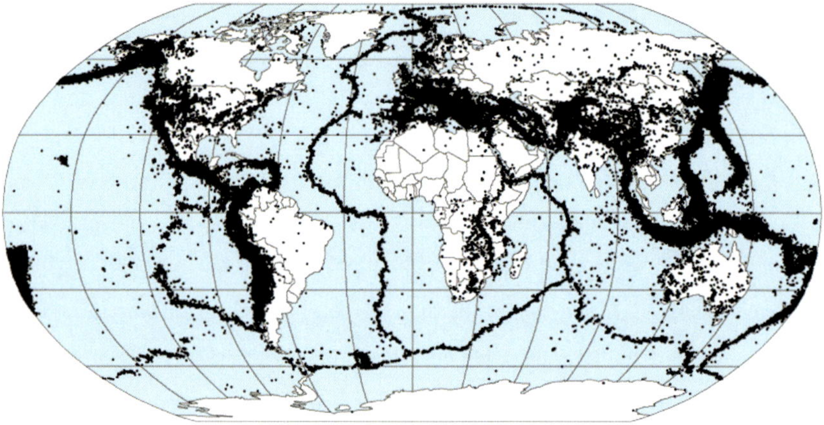

Figure 1.17 Seismicity map of the world in 1963–1998 (a map retrieved from the courtesy of NASA, in Wikimedia Commons).

World seismic activity

In the world, some areas are often subjected to big earthquakes that can be categorized into four major zones as follows:

1. The circum-Pacific seismic zone includes the Pacific side of South America, Central America, North America, the Aleutian Islands, the Kamchatka Peninsula, Japan, Indonesia, and New Zealand.
2. The Eurasian seismic zone extends from Southeast Asia through the Middle East to the Mediterranean Sea.

3. The Mid-Atlantic ridge and the area where it reveals itself on the land.
4. Parts of China, North America, the Middle East, and other continents.

As shown in Figure 1.17, some zones where earthquakes occur frequently are distributed in long, narrow strips. These areas are therefore called seismic zones. Recent large earthquakes of over eight in magnitude have occurred in the trenches of arcs of islands, filling up areas where very few or fewer earthquakes have occurred in recent years. The term seismicity gap is applied to such zones where earthquakes will possibly occur in the near future.

If we observe Figure 1.17, there are many regions or places in which big earthquakes occurred in the past. These regions or places are mostly located at the boundaries of tectonic plates.

Figure 1.18 is a zoning map of the tectonic plates of the world. Currently, the main fifteen tectonic plates are zoned as:

1. African Plate
2. Antarctic Plate
3. Arabian Plate
4. Australian Plate
5. Caribbean Plate
6. Cocos Plate
7. Eurasian Plate
8. Indian Plate

9. Juan de Fuca Plate
10. Nazca Plate
11. North American Plate
12. Pacific Plate
13. Philippine Plate
14. Scotia Plate
15. South American Plate

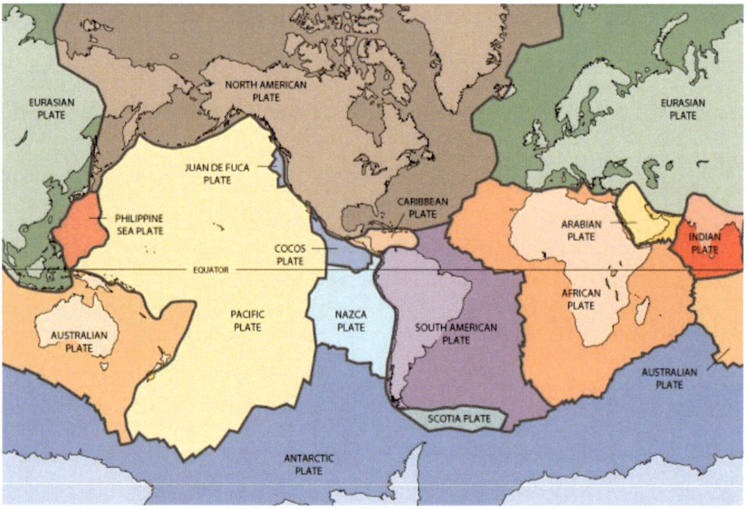

Figure 1.18 Zoning map of tectonic plates of the world (retrieved courtesy of USGS, in Wikimedia Commons).

Among the main plates, there are many smaller sub-plates, which are known as *platelets*. Due to the movements of the main plates, these smaller plates are frequently being thrusted or pinched, which often causes damaging earthquakes with casualties. One example of this occurrence was reported in the 1960 Agadir earthquake in the African continent, and this unexpected earthquake demolished the non-seismic-resistant buildings in the area.

References

[1] Johnston A. & Halchuk S. (1993). The seismicity database for the Global Seismic Hazard Assessment Program. Annali di Geofisica, XXXVI (3–4), 113–151.

[2] National Geophysical Data Center/World Data Service (NGDC/WDS): Significant Earthquake Database. National Geophysical Data Center, NOAA. doi:10.7289/V5TD9V7K

[3] Zamudio Y., Berrocal J. & Fernandes C. (2005). 6th International Symposium on Andean Geodynamics (ISAG 2005, Barcelona), 813–816.

CHAPTER TWO

WHY DOES STRONG SHAKING OCCUR?

Myths and legends of earthquakes

In ancient times, earthquakes were believed to be caused by restless Gods, giant creatures, or animals sleeping underneath the ground. Some myths and legends related to earthquakes in different countries around the world are described below.

In most cases, myths and legends are strongly influenced by supernatural phenomena and a lack of knowledge of the geology of nature. The Greek philosophers developed a theory that earthquakes were caused by the spreading of gases from the ground underneath. Until the eighteenth century, some western scientists, including Isaac Newton, thought earthquakes were caused by explosions of flammable material deep inside the ground. Later, in 1760, the Reverend John Mitchell proposed that earthquakes were mainly caused by the movements of rocks that triggered elastic waves from inside the Earth, which propagated to the surface and caused shaking on the ground.

In ancient Greek mythology, the so-called God of the Sea, Poseidon, was often portrayed carrying a trident. The "Earthshaker" is one of the nicknames of Poseidon. It was believed that the earthquakes were triggered by Poseidon when he got angry by stroking the ground with his trident (Figure 2.1).

Figure 2.1 Poseidon (a sculpture in Copenhagen Port, credited by Hans Anderson, from Wikimedia Commons).

In ancient Japanese mythology, most people believed earthquakes were caused by a monster catfish (*Namazu*) living below the islands and Japan's archipelago. The catfish is controlled by the god *Kashima* who detains the catfish with a sword (Figure 2.2). When the god Kashima loses his harness, the monster *Namazu* struggles, causing big earthquakes in the islands of Japan.

Figure 2.2 The god Kashima controls the monster catfish with his sword, a myth in ancient Japan (an illustration from International Research Center for Japanese Studies, Edo 1855, from Wikimedia Commons).

In Hindu mythology, the old stories from China and India show several elephants standing on the back of a turtle, which trembles and shakes the Earth when any of these animals move (Figure 2.3).

Figure 2.3 An 1876 drawing of the world supported on the backs of four elephants, themselves resting on the back of a turtle (an illustration from Popular Science Monthly Volume 10, 1876, from Wikimedia Commons).

Mythologically, Native American people in the Pacific Northwest believe that the fight between a large thunderbird and a whale in the sea was the cause of the ground shaking and flooding (Figure 2.4). Currently, we call this flooding a tsunami.

Figure 2.4 The Thunderbird with the Whale in its claws (a photo credited by Joshua Murdough, 2009, from Wikimedia Commons).

The structure of our Earth

The inside structure of the Earth consists of the following layers: an outer silicate solid crust, a viscous asthenosphere, a dense and rigid mantle, a stiffer mantel of liquid outer core that is less dense than the rigid mantle, and a solid inner core (Figure 2.5). Scientific understanding of the Earth's internal structure is based on observations of topography and bathymetry, interpretations of rock in outcrops, samples brought to the surface from greater depths by volcanoes or volcanic activity, and analysis of the seismic waves. Measurements of the gravitational and magnetic fields of the Earth and experiments with crystalline solids at pressures and temperatures characteristic of the Earth's deep interior are discussed below.

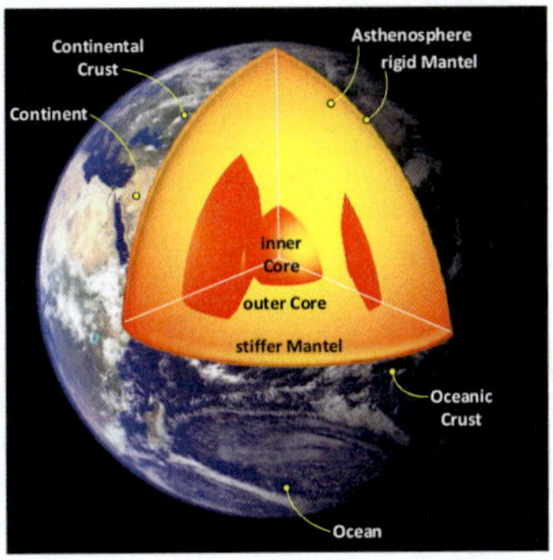

Figure 2.5 The illustrative structure of the Earth.

The crust

The crust is the outermost brittle layer that varies in thickness from 25 to 70 km, where it forms most continents or 5 to 10 km under the sea. The continental crust is less dense and is composed of sodium potassium aluminum silicate rocks like granite. The oceanic crust is composed of dense iron magnesium silicate igneous rocks like basalt.

The mantle

The mantle is the thickest layer in the structure of the Earth. The mantle layer is about 2,890 km below the crust layer that comprises dense silicate rocks. Both the longitudinal P- and transversal S-waves from earthquakes are transmitted through the mantle, indicating that the mantle is a solid layer (Figure 2.6).

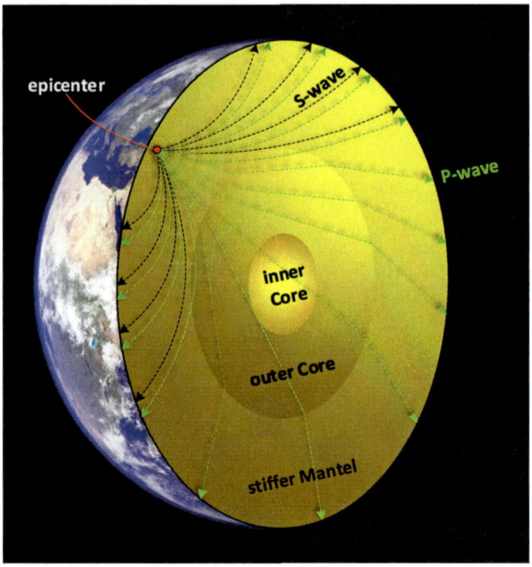

Figure 2.6 Visualization of P- and S-waves transmitted through the Earth's interior to the surface.

The core

Seismic measurements show the Earth's core can be divided into two parts: a solid inner core with a radius of about 1,220 km; and a liquid outer core with a radius of about 3,400 km.

The inner core is composed primarily of iron and some nickel. Because the layer can transmit transversal S-waves, it should be a solid of metal. The outer core surrounding the inner core is believed to be composed of iron mixed with nickel and lighter elements. Because the outer core does not transmit transversal S-waves, it should be a metal liquid.

What is a tectonic plate?

In 1915, the theory of continental drift was pioneered by a German meteorologist, Alfred Wegener, to explain the phenomena of earthquakes. Plate tectonic theory became one of the principles to explain the earthquake mechanism due to continental drift.

The theory that Wegener construed was that about 200–300 million years ago, a single land mass slowly broke up and drifted apart to form continents. Wegener pointed out that the separate continents looked like they were broken up along the coastlines from a single land mass (Figure 2.7).

Figure 2.7 Illustration of the closed and open Atlantic Ocean.

It was 50 years later that the continental drift theory was recognized. The fundamental principle of the continental drift theory is that the Earth's crust consists of several plates that move by floating on the rigid mantle. The movements of the plates are extraordinarily slow but produce tremendous forces at the plate boundaries. Plate movements are governed by the thermomechanical equilibrium of the mass inside the Earth. The rigid mantle is in contact with the cooler crust, while the stiffer mantle is in contact with the hot outer core. A temperature difference occurs within the mantle, resulting in the cooler and dense crust laying on the less dense but warmer mantle. These conditions are believed to drive the movements of plates. The cooler, denser crust will subduct into the mantle, and the less dense warmer mantle will rise, repeating this cycle. The movement of continents can be noticed by observing similar plant and animal fossils found along the coastlines of different continents, which indicates that the continents were joined together in the past. Aided by modern recording and measuring instruments, the sea floors can be seen to spread apart in some locations where molten rock is emitted between the continents. The continental drift theory hence gained recognition in the scientific community. It was seen as one of the most revolutionary ideas in geology from which the discipline of plate tectonics began.

As well as the major tectonic plates, the many smaller plates are known as platelets. These platelets move due to the movement of the tectonic plates. The platelets can cause damaging earthquakes that may occur almost anywhere in the world. These unpredicted earthquakes can be destructive, like the 1960 Agadir earthquake in Africa, where almost none of the buildings in the affected area had been designed to have earthquake seismic resistance.

What is a fault?

There are three types of connecting boundaries due to the movements of tectonic plates (Figure 2.8).

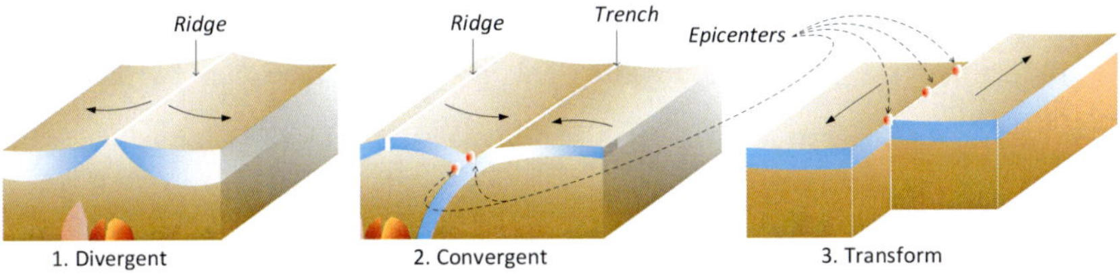

Figure 2.8 Illustration of three types of tectonic plate boundaries.

They are as follows:

1. A divergent boundary is where two plates move apart. A new lithosphere layer is produced, or the old layer becomes thinner. When both plates move away from the boundary, it is known as Spreading Centers. Examples of divergent boundaries of plates are the mid-oceanic ridges.
2. A convergent boundary is where two plates move toward each other and collide. The lithosphere layer between the two plates becomes thick or sinks into the mantle. Examples of convergent boundaries of plates are the subduction zones and alpine belts.
3. A transform boundary is a boundary where two plates move toward each other without making any collisions. Examples of transcurrent boundaries are the transformed faults and other strike-slip faults.

The divergent and transform boundaries with hypocenter depths less than 30 km are associated with shallow earthquakes. Subduction zones and regions of continental collision can have a much deeper hypocenter or focal depth [1].

Faults are formed in small-scale areas, mostly after earthquakes and usually near the ground surface. The faults that emerge on the ground surface because of an earthquake are called earthquake faults. Earthquake faults are not formed by deep earthquakes. The ground surface can be stressed, strained, or sheared during the earthquake, thus undergoing slippages. Depending upon direction, the slippages are classified as follows:

1. Dip-Slip. A kind of slippage that takes place in a vertical direction.

a. Normal fault. The left side ground layer slips downward (Figure 2.9a).

b. Reverse fault. The left side ground layer slips upward (Figure 2.9b).

2. Strike-Slip. A kind of slippage that takes place in a horizontal direction.

a. Left lateral fault. The left side ground layer slips toward the front left (Figure 2.9c).

b. Right lateral fault. The left side ground layer slips toward the back left (Figure 2.9d).

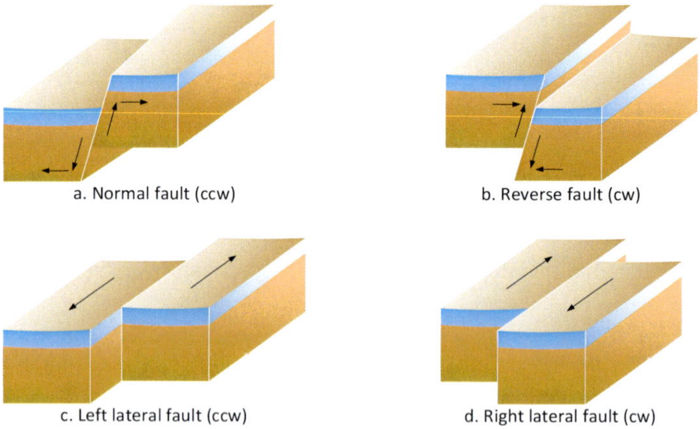

Figure 2.9 Illustrations of fault motions.

The actual faults formed after earthquakes often combine the four types of slippages. The best-known example of an earthquake fault is the 1,300-km-long strike-slip of 8.5 m at the San Andreas fault, which caused the San Francisco earthquake of 1906 (Richter magnitude of 7.9). During the El Centro earthquake of 1940 (M = 6.9), a 60-km-long right lateral fault was created with a maximum slip of 4.5 m.

One of the most famous faults in Japan was created by the Mino-Owari earthquake (M = 8.0) in 1891 (Figure 2.10). It is 80 km long and displays a 2 to 3 m vertical slip and a maximum 8 m horizontal slip.

China's Haiyuan earthquake of 1920 (M = 7.8) created a left lateral fault 200 km long.

Figure 2.10 The Neodani normal fault in Motosu, Gifu, Japan, formed after the Mino-Owari earthquake in 1891 (a photo credited by Tomomarusan in Wikimedia Commons).

When an earthquake fault is created by relatively small earthquakes of M less than 6, the length and width of an earthquake fault are usually correlated. Earthquake faults are longer when earthquakes are greater.

Faults are initially caused by big earthquakes. However, the faults can also trigger an earthquake as follows:

1. The accumulated strains and stresses stored in the fault for a long time start relieving (Figure 2.11a).

2. Slip occurs at fault and causes a rebound (Figure 2.11b).

3. A push-and-pull force acts at fault (Figure 2.11c).

4. Two pairs of couples with equivalent forces suddenly acting at fault (Figure 2.11d).

5. The fault triggers an earthquake by transmitting radial or P-wave seismic propagation to the surrounding area.

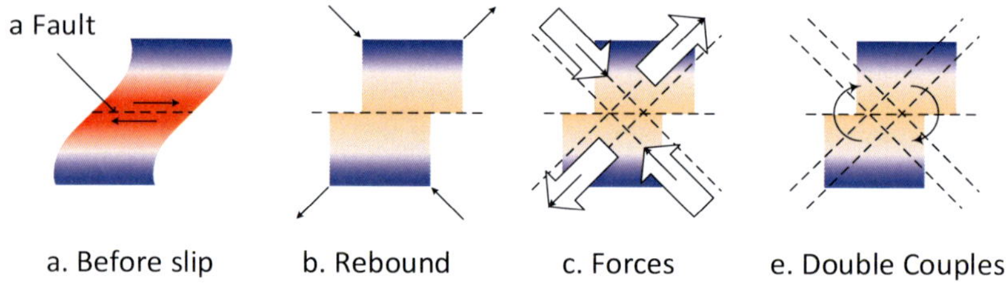

Figure 2.11 Illustrations of the potential fault that can trigger an earthquake.

The moment of each couple is called the earthquake or seismic moment [2]. The seismic moment is defined as the rigidity of the ground× the area of faulting× the amount of the slip. Recently it has been used as a unit measure of an earthquake or seismic moment.

Active faults have undergone deformation for several hundred thousand years and could potentially trigger an earthquake. Geological and topographical surveys and aerial photographs have found some active faults. When designing a vital structure such as a nuclear power plant to resist seismic forces, the distance between nearby faults and the building site must be considered in predicting the ground's motion.

The famous San Andreas Fault in California is far from Point Arena and the Gulf of California. A right lateral fault was formed where the Pacific plate slips to the north against the North American plate. It caused the Fort Tejon earthquake (1857), the San Francisco Earthquake (1906 and 1957), the Loma Prieta earthquake (1989), and the Parkfield earthquake (2004).

Average slip velocity at an active fault varies. The highest velocities of the San Andreas fault and the Nankai trough of Japan are 10 to 100 mm/year. Therefore, a slippage of 3 m during one earthquake means that earthquakes occur every 30 to 100 years at these faults. Some active faults, such as the San Andreas fault, are always moving; others, such as some in Japan, move only when an earthquake occurs.

What is a hypocenter?

The source of an earthquake inside the crust is called the earthquake's hypocenter. The *epicenter* is a point on the Earth's surface vertically straight above the *hypocenter*. The epicentral distance is the horizontal distance between a location with the epicenter. The distance from the focus or hypocenter to the location is called the focal distance. The distance between the focus and the hypocenter is called the focal depth.

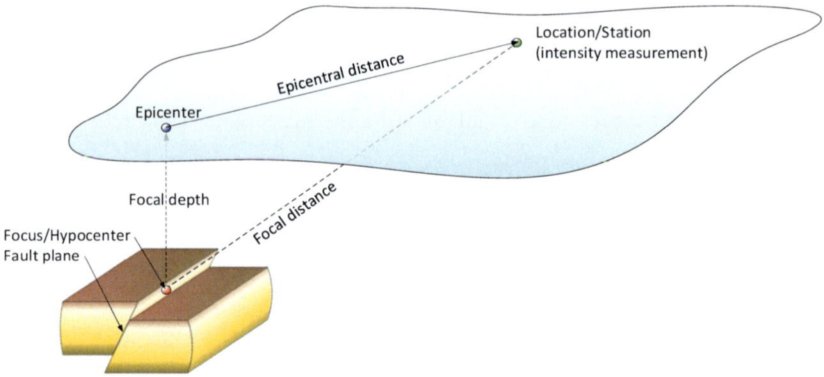

Figure 2.12 Descriptions of focus and hypocenter of an earthquake.

Seismic waves propagate from the focus through a certain region of the surrounding ground, which is called the focal region. Earthquakes can be categorized as shallow (less than 70 km), intermediate, and deep (beyond 300 km), depending on the depths of their foci from the ground surface.

References

[1] Lay, T. and Wallace, T.C. (1995). Modern Global Seismology, Vol. 58, International Geophysics Series. Academic Press, San Diego.

[2] Kasahara, K. (1981). Earthquake Mechanics, Cambridge University Press, London.

CHAPTER THREE

WHAT ARE THE ROLES OF THE SOIL AND GROUND?

Soil and Ground that change the seismic wave

The inside of our Earth is composed of many substances, each of which has different physical characteristics. Seismic waves change with the different physical properties each time they pass through boundaries. To investigate the characteristics of seismic wave propagation, we must understand how different underground substances affect seismic waves.

Figure 3.1 shows the composition of the Earth at various depths. On a very large global scale (Figure 3.1a), the Earth is divided into the crust, mantle, outer core, and inner core. The most damaging earthquakes occur in the crust and the upper part of the mantle. The structure from the surface to several tens of kilometers underground is closely related to the ground's motion generation and propagation characteristics. This depth is only about 1/200th of the Earth's radius of about 6,400 km. If the Earth is visualized as an egg, it corresponds only to the shell thickness. Figure 3.1b shows the nonuniform crust sedimentary layers, separated into upper and lower parts by the *CONRAD* discontinuity. The boundary between the crust and the mantle is known as the *MOHO* (Mohorovičić) discontinuity. In addition, the thickness of the crust varies from place to place. In the sea areas, the crust is thin, and in the land areas, it is as thick as about 30 km (Figure 3.1b).

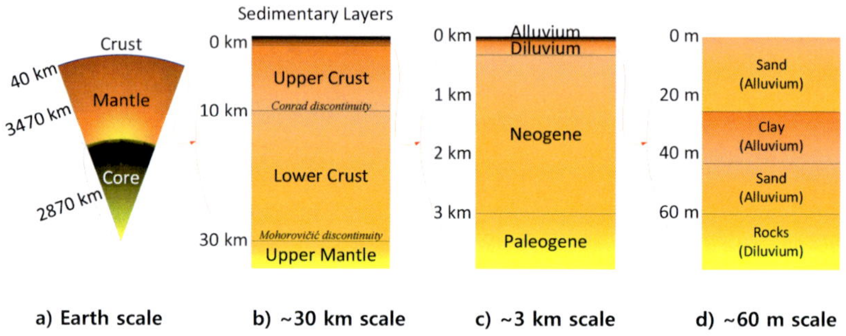

Figure 3.1 Earth structure shown on different scales

Furthermore, in a large plain, there is a sedimentary layer with a thickness of 2 to 3 km above the top layer of the crust (Figure 3.1c). In terms of geological age, this sedimentary layer was deposited from 2.5 million to 200,000 years ago, called the Neogene period, compared to the Earth's history of 4.6 billion years. It was the result of a new era of natural phenomena.

These sedimentary layers are called "deep ground." As they get deeper, the density and the speed at which seismic waves propagate increase due to the effects of consolidation. Even in the same plain as in coastal areas, there is a "Quaternary Formation" deposited in a newer era (from about 200,000 years ago to the present) above the deep ground (Figure 3.1d). Furthermore, the Quaternary layer is divided into the relatively hard "Diluvium" and the softer "Alluvium" formed in the newest era. The alluvium is found along the coast and along rivers. The alluvium can be as thick as tens of meters. These strata are collectively called the "bedding surface."

Since the traveling speed of the S-wave (technical detail described in Chapter Four) is almost proportional to the stratum's square root of the hardness (shear modulus), the harder the stratum, the higher the S-wave velocity. Also, the softer the ground, the more the ground motion amplifies.

Regarding S-wave velocities, the changes in S-wave velocities in the crust and mantle are summarized in Table 3.1.

Table 3.1 Approximation of S-wave values.

Stratum		**S-wave velocity (km/s)**
	Sand	0.1 0.2
Alluvium	Clay	0.2 ~ 0.4
	Bedrock	0.3 ~ 0.5
Diluvium		0.8
Neogene		1.5
Paleogene		3.0
Upper Crust		3.5
Lower Crust		3.8
Mantle		4.4

Research on the crust and mantle has been conducted for a long time in seismology. It is still an important research theme that is the basis for understanding seismic phenomena. On the other hand, in soil engineering and geotechnical research dealing with bedding surfaces, it is rare to consider the propagation of seismic waves in the structure of the crust and mantle. Ground motion seismology is an interdisciplinary field that connects the two. The underground structure to be handled is wide, from the crust and mantle to the bedding surface.

Changes in seismic waves from the hypocenter

The interior of the Earth is composed of many substances, seismic waves generated by shallow earthquakes propagate through the crust, and seismic waves from deep earthquakes propagate through the crust and mantle. In addition, seismic waves reach the surface through deep ground and bedding. It is easier to understand the changes in seismic waves between the crust and mantle and the deep ground and bedding surface. The former is called the "propagation path" characteristic, and the latter is called the "ground" characteristic. The propagation path and the ground are still underground strata. Still, the idea of "basement" has been introduced to distinguish between the two. The upper surface of the basement is set at the boundary surface, where the impedance (product of velocity and density), determined for each stratum, changes considerably. Therefore, in the definition of the basement, we have to consider the change of S-wave velocity and density underground.

Until the early 1970s, the basement often represented the upper surface of the stratum with an S-wave velocity of several hundred meters per second. This stratum serves as a support layer when constructing relatively tall buildings. However, with economic growth and the construction of many large-scale structures such as skyscrapers and long bridges, the seismic waves in the wide and deep ground have been used to evaluate these structures' safety. The amplification effect was also important, and it was sometimes necessary to set deeper foundations. In addition, the term infrastructure is used in various fields, such as Earth science and seismic engineering. Therefore, the bedrock with an S-wave velocity of about 3 km/sec is called the "engineering earthquake base" in particular. If the earthquake base is used, the amplification effect of seismic waves on the ground above the earthquake base can be treated as a ground characteristic, as shown in Figure 3.2. An earthquake base at about 2.5 km in an urban area is almost near the foot of a mountainous area. The stratum underground at all points in a large area, such as a large plain, is also an important feature of the earthquake base.

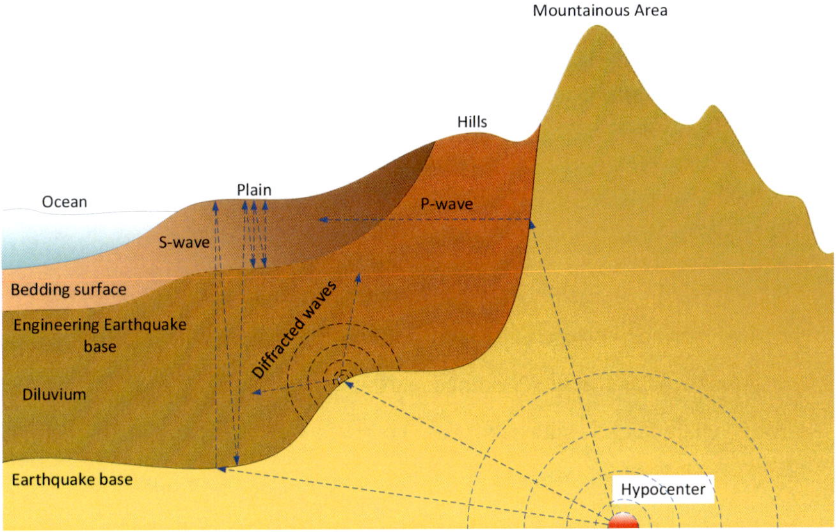

Figure 3.2 Schemes of diffracted waves through different soil types.

It is more convenient to use S-wave velocity to evaluate the ground motion rather than the geological characteristics. Normally, the stratum with a three-wave velocity of about 0.4 km/s is called the "etiological base." The stratum above it is the surface stratum. This engineering earthquake base is a stratum that serves as a support layer for general structures.

In the crust and mantle, the change in seismic wave velocity is relatively small for an underground structure deeper than the earthquake base. The change due to the reflection and refraction of seismic waves generated from the hypocenter is not large, as mentioned above. Therefore, as will be described later, only the effect of reducing the amplitude according to the distance is often considered in the propagation path.

There are three characteristics: radiation characteristics of seismic waves at the hypocenter (earthquake source characteristics), attenuation characteristics of seismic waves at the propagation path (propagation path characteristics), and amplification characteristics at the deep ground and bedding surface (ground characteristics). Considering the effects individually and superimposing them at the end, the ground motion observed on the bedding surface can be obtained. According to this idea, ground motion on the bedding surface can be calculated by finding the seismic motion at the earthquake base for the assumed earthquake and multiplying it by the ground characteristics at each point. This idea is very convenient when you want to know the distribution of the strength of ground motion in a certain area, such as the earthquake damage estimation of a local government.

In the occurrence of an earthquake, the seismic waves change their propagation paths across boundaries of different media. In general, seismic waves radiated from a hypocenter decrease in amplitude as the distance increases. Since the change in the velocity of seismic waves in the propagation path is smaller than that in the ground, consider it a homogeneous medium with no boundary surface. A seismic wave propagates through the medium, and the amplitude of the seismic wave decreases with the increase in distance. This phenomenon is due to the following multiple effects. One effect (geometric attenuation) is that the amplitude becomes smaller as the seismic wave spreads from the hypocenter. The other is the effect of reducing the amplitude (viscous damping and scattering damping) because the ground is not homogeneous. These effects are collectively called "distance attenuation."

Geometrical attenuation occurs when the amplitude of the body wave (S-wave or P-wave) decreases by increasing the distance from the epicenter. On the other hand, the amplitude of the surface wave attenuates in proportion to the square root of the epicentral distance. Therefore, the surface waves have a smaller effect on distance attenuation. The farther away the location, the more dominant the surface wave is than the original wave. Since the surface wave propagates along the Earth's surface, the amplitude's attenuation follows the epicentral distance.

The ground changes with strong shaking

The theory of elasticity underlying the propagation of seismic waves is the theory of small deformation. However, when a strong seismic wave occurs, the ground deformation undergoes a large deformation. Hence, the modulus of rigidity of the ground decreases. These effects are called "non-linear effects of ground motion." A decrease in modulus greatly reduces the shaking due to an earthquake. The distortion is proportional to the amplitude of the seismic wave divided by the wavelength. The distortion is small because the amplitude and the wavelength are long in deep ground. Therefore, in deep ground, non-linear effects can be ignored. However, on a soft bedding surface, the amplitude is large, and the wavelength is short, so the strain level is large, and the influence is significant. In addition, the ground "liquefies" in the sandy ground containing water during strong earthquakes. The liquefaction of the ground was noted as an apartment complex with reinforced concrete structures in Kawashi town, Niigata City, that fell over without major loss of structure during the Niigata Earthquake of 1964 (Figure 3.3).

Figure 3.3 Liquefaction during Niigata Earthquake 1964 (Japan National Committee on Earthquake Engineering, Proceedings of the 3rd World Conference in Earthquake Engineering, Volume III, pp 78–105).

Since then, liquefaction has been actively researched, mainly in soil engineering. Liquefaction is a phenomenon where the water pressure inside the pores in sand particles increases due to earthquake shakes (Figure 3.4).

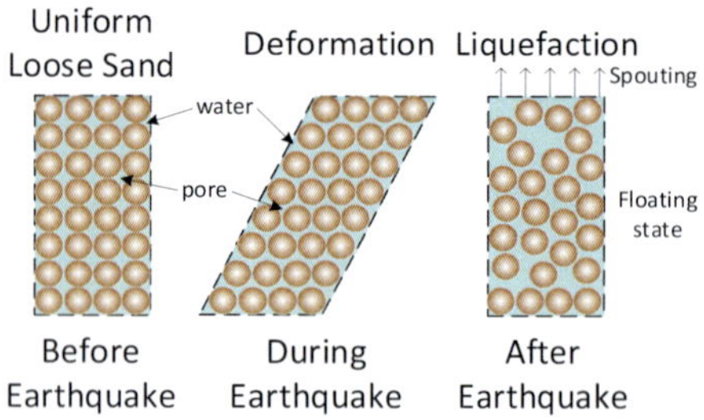

Figure 3.4. Mechanisms of liquefaction during an earthquake.

When the water pressure increases, the frictional forces between the particles reduce, and the sand is in a state identical to a liquid. Therefore, water and sand move up to the surface, and the ground and heavy objects sink or fall down.

Amplification of seismic waves in the ground

The seismic damage distribution is closely related to the thickness of the alluvium, which is the sedimentary layer. The seismic intensity and damage distribution can often be roughly explained by the thickness of the alluvium. Scientific studies based on seismic observation data are indispensable for explaining the mechanism of ground amplification and when seismic observations are being carried out under various ground conditions.

Figure 3.5 shows ground motion records obtained in and around the rock above and below the ground. Seismic waves in the surroundings and bedrock have almost the same characteristics, having grown a little and having a long duration. At points on the surface of the ground where the soft sedimentary layer is thick, the amplitude is even larger, and the duration is longer. There is an intermediate feature between the two seismic waves at the point where the bedrock and the sedimentary layer are thick and thin on the surface ground.

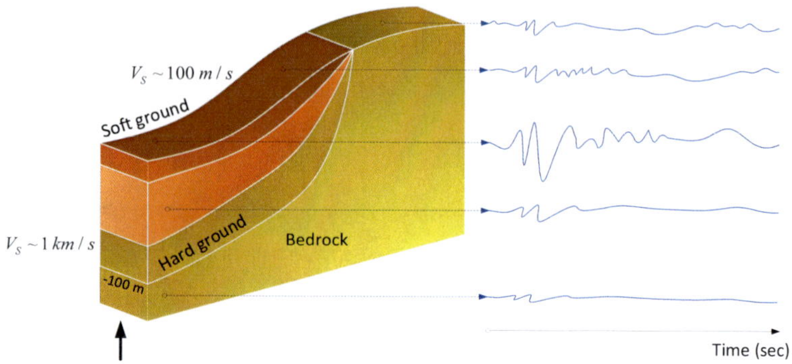

Figure 3.5 Amplification of seismic waves across the soil boundaries.

In Figure 3.5, the periodic components also differ depending on the observation point. In other words, at the point where the sedimentary layer is thin, it shakes in a short cycle, and at the point where it is thick, it shakes slowly and in a long cycle. This shaking cycle is unique to the sedimentary ground. It is a phenomenon seen in most seismic records on the ground. This period is called the "dominant period," and its periodic component is amplified most, so it is one of the most basic characteristics of the ground characteristics.

Earthquake waves are composed of various periodic components, making the ground either liable to shake or difficult to shake. This, which shows the amplification rate for each cycle, is called the "amplification rate spectrum."

As shown in Figure 3.5, if a sedimentary plain is surrounded by bedrock, surface waves from the hypocenter will hit the boundary between the plain and the mountains. Then, a reflected or transmitted surface wave similar to a body wave is also generated at this boundary surface. In this case, the boundary surface is only the thickness of the sedimentary layer, so it is strictly different from the case of the body wave. However, since the surface wave transmitted to the plain side propagates to the smaller impedance side, the amplitude becomes larger similarly.

Surface waves that enter directly cause a phenomenon called "dispersion" that is not found in real waves due to the influence of the sedimentary layer structure. Dispersity is a phenomenon in which the propagating speed differs depending on the period of the seismic wave, which is unusual in surface waves. The longer the periodic component, the higher the propagation speed; the shorter the period, the lower the propagation speed. As a result, the longer the distance, the longer the period of seismic waves will advance, and the greater the difference in the arrival times of surface waves of different periods. Hence, the longer the seismic tremors will continue.

Further, as described above, in the case of surface waves, the small geometric attenuation also causes a long duration. Also, suppose there is a boundary with a similar bedrock on the other side of the plain. In that case, the surface wave will be reflected and return to the original direction. This is a horizontal version of the overlapping reflection phenomenon of S-waves, and the shaking in the plain will continue for longer.

The actual deposition plain is not two-dimensional underground but a basin with a three-dimensional region, and the propagation of surface waves becomes more complicated. The earthquake base is shallow at the edges of the plain, and the sedimentary layer is generally the thickest near the center. As mentioned above, the propagation velocity of surface waves differs depending on the period and the underground structure's composition.

Even with surface waves of the same period, the propagation velocity is low when the soft stratum is thick. Conversely, it is high when the soft stratum is thin. In a basin where the thickness of the sedimentary layers changes, the propagation velocity of surface waves also changes. The surface waves on the plain do not travel straight but propagate while bending according to changes in the underground structure. The underground structure affects the propagation of long-period surface waves. The surface waves transmitted through multiple paths overlap, resulting in large amplitudes and long durations.

Mechanisms of long-period shaking

Long-period shaking is believed to be due to the surface waves of a shallow terrestrial earthquake. Even though the earthquake may have occurred far away, this phenomenon of long-period shaking is common on large plains. Of course, the long natural period of high-rise buildings is also a reason, but this will not happen without the seismic wave motion containing long-period components.

Thus, long-period seismic motion affects high-rise buildings and other large, slim structures. Bridges and towers with long spans are also prone to shaking due to long-period seismic motion. In addition, the liquid in a large tank reacts sensitively to long-period shaking, and the liquid level may shake significantly. This was one of the causes of the fire in the large oil tank in the mid-1970s, and not much research had been carried out on these seismic motions within several seconds. The period of seismic motion, which was shorter than one to two seconds, attracted attention in seismic engineering because the natural period of many structures to be studied was around that level.

On the other hand, in seismology, the main research subject using seismic waves is the Earth's internal structure deeper than the upper crust. Less research has been conducted on seismic waves with a period shorter than about 10 seconds, which is affected by the sedimentary layer. However, with rapid economic growth, large-scale structures such as high-rise buildings have begun to be built in metropolitan areas. The seismic design of such large-scale structures (Figure 3.6) required scientific research on the long-period seismic motion mentioned above.

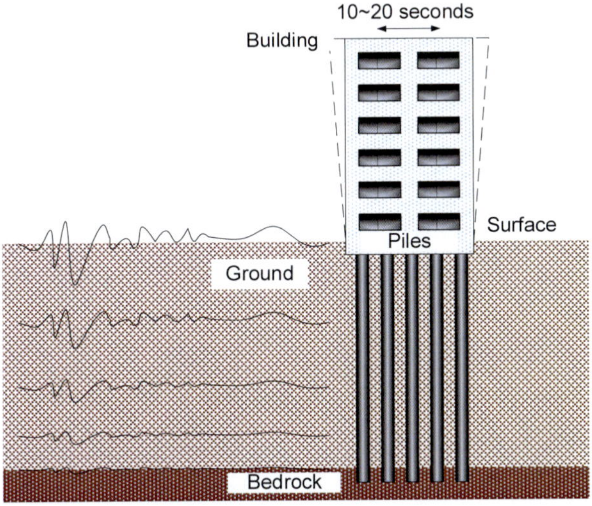

Figure 3.6 A high-rise building experiencing long-period shaking.

Mathematical modeling of soil and ground

Along with the rapid acceleration of modern computers, the use of computational mechanics in research trends is massively progressing. The

Finite Element Method (FEM) has become the most popular numerical method for scientists, researchers, engineers, and practitioners. In FEM, the actual physical system under study is idealized into a mathematical model. The model is then discretized into typical elements using the finite element concept (Figure 3.7).

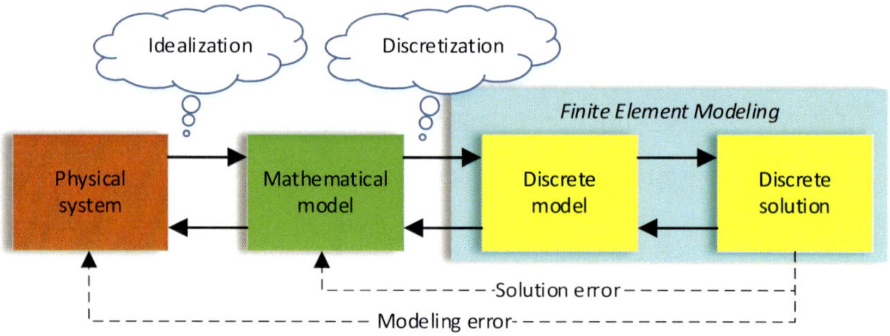

Figure 3.7 Mathematical modeling of a physical system in nature.

The solutions of theories behind these elements are derived and interpolated at all the element's nodes, usually distributed along the boundary perimeter of an element.

One good example to show a physical system to elucidate the concept of FEM is a harp, as shown in Figure 3.8.

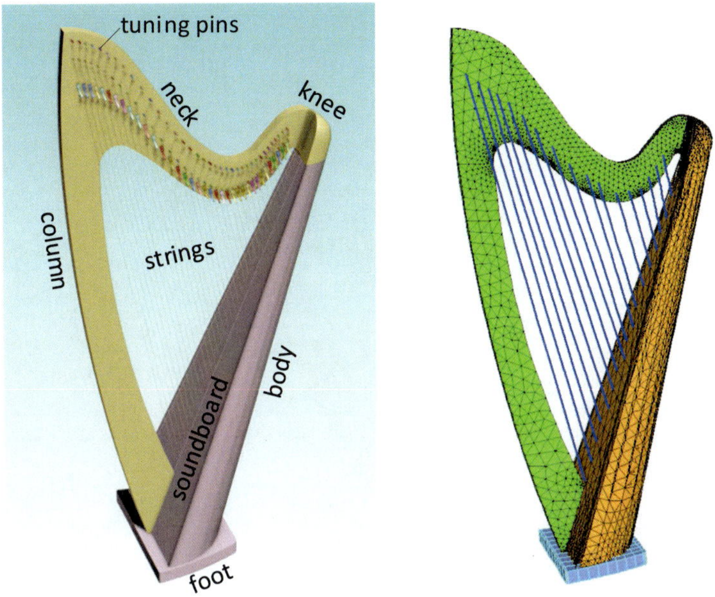

Figure 3.8 FEM concept by using a harp [1].

A harp is a musical instrument with individual strings mounted at an angle to its closed V-shaped wooden frame. To produce the sound, the strings are plucked with the fingers. Harps have been recognized since antiquity in Asia, Africa, and Europe, dating back at least as early as 3,500 BC. The instrument had a great reputation in Europe during the Middle Ages. Renaissance evolved into several variants with new technologies and was disseminated to Europe's colonies. The instrument also found a particular reputation in Latin America. Although some ancient members of the harp family died out in the Near East and South Asia, descendants of early harps are still played in Myanmar and parts of Africa. Musicians in the modern era still utilize other obsolete variants in Europe and Asia.

Figure 3.9 shows a harp modeled using different elements (line, plane, block).

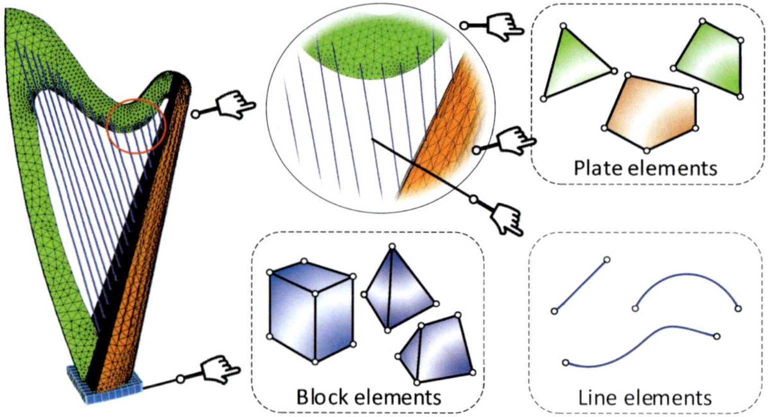

Figure 3.9 Representing a harp using different elements [1].

Modeling a real physical system requires the modeler's engineering judgment to choose the preferred element type. In fact, the results of prediction and solutions errors depend on the type of element used in the model.

The elements shown in Figure 3.10 can be assembled across the different types to build a model by connecting their nodes to their adjacent element(s). Under a defined "rule of connection" given to a node, the node can act as a connecting terminal between two or more elements with similar rules.

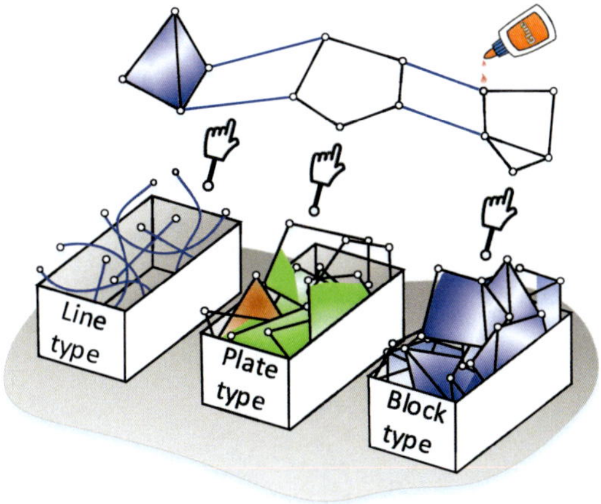

Figure 3.10 Assembling elements to build a harp model [1].

The selection of element types for FEM modeling depends on the nature of the problems. Most element types used in a structure are line, plate, and block. Variations can be added to these element types by adding a node or using curvatures at the boundaries. Plenty of rules govern the element depending on the theories assumed and the type of elements used in a model.

A plate-like element is used to model soils in 2D problems with an infinite thickness relative to its on-plane dimensions (so-called "plane strain" element), or an axisymmetric plate-like element is used. In 3D problems, brick elements with infinite boundaries are widely used in practice. To model buildings in 2D and 3D problems, line elements, plate-like "plane stress" (when an element has a thickness), and brick elements are usually used in practice and research.

Earthquake simulation of soil and building

Earthquake simulations of soil and buildings using computers are often performed using FEM. Because of the recent improvements in computer performance, even a personal computer can finish the calculation instantly. Hence, the simulation of seismic waves propagating through the entire deposited plain has become possible. Figure 3.11 shows an example of earthquake simulation using a personal computer.

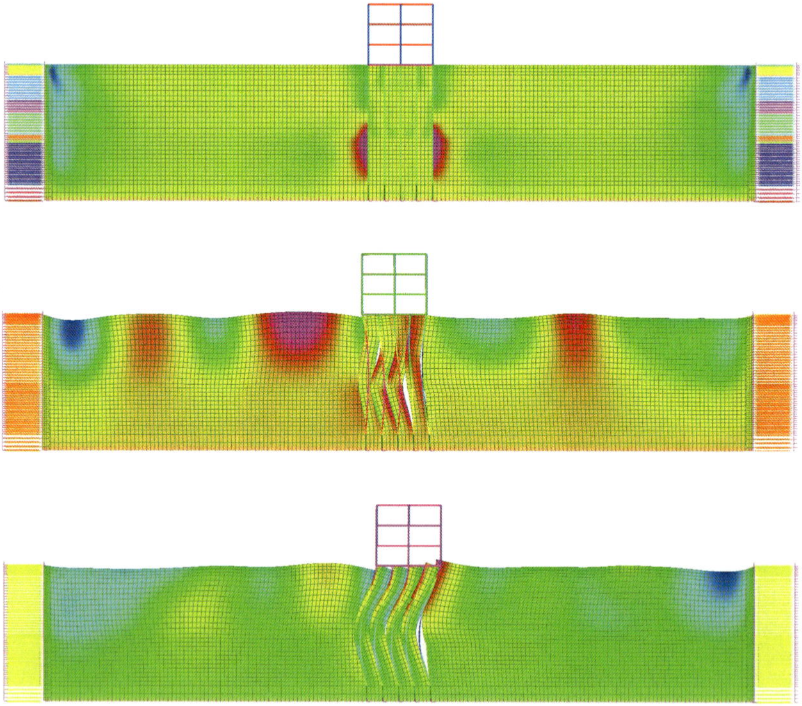

Figure 3.11 Example of full-scale earthquake simulation of a building.

Several approaches can be considered in modeling soil and buildings for an earthquake simulation. As shown in Figure 3.11, the building supported by pile foundations is modeled with the surrounding soils and bed rock base where the seismic wave enters. The figures show three different selected conditions during a full-time simulation. We can notice gaps between the pile foundations and adjacent soils because the spring (not shown) is set to be inactive when the movements of the soil and piles are separating from each other. On the contrary, the spring becomes inactive when the soil and piles pull together.

On the soil's left and right ends, springs and dampers are installed, representing a "semi-infinite" condition of the soil, where the ground is continuously spread to an infinite length. These "semi-infinite" elements have special features to absorb the incoming waves during the simulation, not to be bounced back as in the real situation where the waves will fade away.

As a result of the simulation, the building model can be checked for its strength, and the pile foundations can be examined for their capacity to support the weight of the building to satisfy the required design codes.

Although the example is in a two-dimensional model, it is considered comprehensive enough to simulate a building during an earthquake.

Earthquake simulation as a tool

Planning and design are the most important stages in building engineering before construction begins. Simulating real building behaviors against earthquakes is only possible through a real-scale experiment or test. However, such an experiment requires enormous equipment and facilities, which are prohibitively costly. In contrast, earthquake simulation is the best alternative to pursue. Research in soil and building behaviors during earthquakes has been intensively conducted for a few decades, yielding accurate predictions without any experimental work.

To list a few advantages of using computers to do earthquake simulations can give the following:

1. Economical design of buildings/houses by finding the optimal solutions.
2. The appropriate foundation type is necessary for parametric modeling to obtain the best results (Figure 3.12).

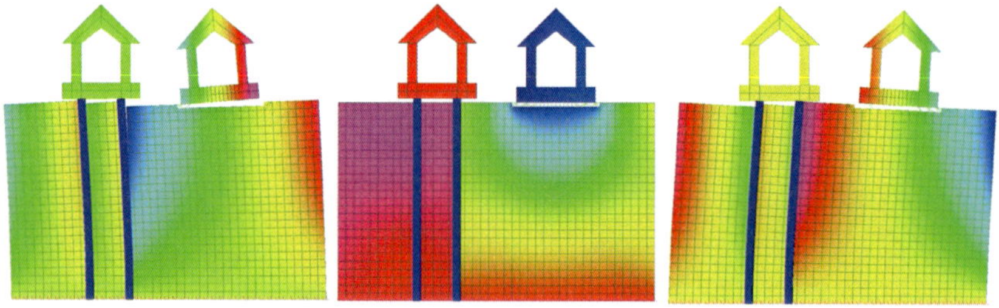

Figure 3.12 Earthquake simulation of investigating the foundations of a house.

3. An efficient tool for evaluating the construction methods available.
4. The output of research and development of new seismic devices.

5. Measuring performances of newly developed seismic devices without doing real experiments to verify.

6. Evaluation of cost-benefit analyses of construction method performances (Figure 3.13).

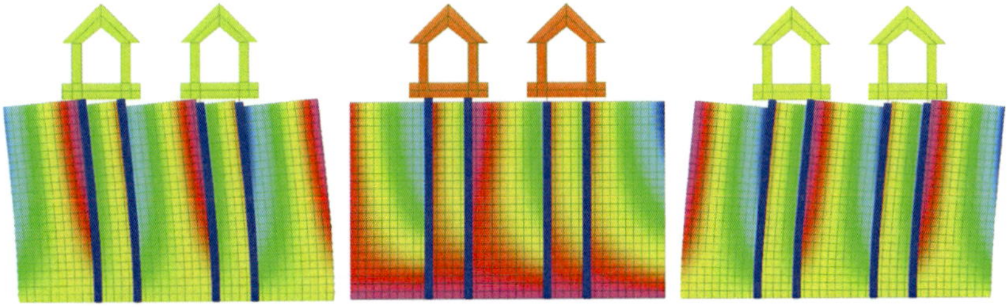

Figure 3.13 Earthquake simulation investigating pile performances.

It is worth noting that conducting experiments on-site will be costly and time-consuming at the stage of the design and evaluation of a building. Therefore, simulation has been widely considered a solution to study and evaluate the feasibility and economic benefit of pre-designing a building.

Reference

[1] Buntara Sthenly Gan (2019). Computational Modeling of Tensegrity Structures: Art, Nature, Mechanical and Biological Systems, Springer.

CHAPTER FOUR

A NEW PHILOSOPHY IN SEISMIC DESIGN

14:46 on March 11, 2011

An earthquake occurred around 14:46 on March 11, 2011, off the Pacific coast of Tōhoku. The earthquake's Richter magnitude was 9.0, the largest earthquake in the history of observation around Japan at the time of its occurrence. On that day, the author was in a classroom on the fifth floor of building No.55 at the College of Engineering, Nihon University Koriyama Campus in Fukushima. The author still remembers the sudden shaking of the floor horizontally with a very large sway. On the night of the same day, the news on TV caught the author's attention, saying that the city's seismic intensity was less than 6 (5-strong). It was the first experience for the author to feel such a long and strong shaking. The duration of the earthquake at that time was 160 seconds. It was a long time shaking on the fifth floor of a building.

Ten years later, an earthquake with a magnitude of 7.3 with a duration of 17.5 seconds occurred at 23:07 on February 13, 2021. It was considered an aftershock of the 2011 earthquake because it was close to the previous hypocenter. Why was a seismic intensity of 6-weak observed in the city despite the different magnitudes of the two earthquakes? The differences in seismic intensities reported were due to the differences in epicentral distances of the cities to the epicenters and the hypocenters' focal depth.

According to the database, about 20% of earthquakes with a magnitude of six or more occurred in Japan. Hence Japan is said to be an earthquake-prone country. The Japanese archipelago is where the Earth's four large plates meet. Since many plates of the Earth are moving in different directions at speeds of several centimeters a year, the boundaries between the plates break, causing earthquakes that shake the Earth's surface.

The word "earthquake" refers to the phenomenon where the ground surface trembles. In other words, it shakes. However, the destruction of plates that occurs in the ground that causes shaking has also been called an earthquake. The earthquake, in this case, refers to the hypocenter. What is the magnitude announced when an earthquake strikes? A *Magnitude* is a unit that indicates the amount of energy caused by an earthquake. It is represented by the letter "M." Invented by American seismologist Charles Richter, experts sometimes refer to it as "Richter magnitude."

The relationship between magnitude and seismic intensity is likened to a light bulb's power and a room's brightness. A bright light bulb can illuminate a long distance. It is bright right under the light bulb, but it gets a little darker away from it. Magnitude refers to the amount of energy generated from the hypocenter, and seismic intensity indicates the degree of shaking at a specific location. Before 1995, the strength of seismic motion was determined based on the perception of the human body, the shaking of surrounding objects, and the degree of damage. In Japan, when an earthquake is felt by the human body, the JMA [1] reports the degree of shaking. The seismic intensity given by the JMA shows the unique seismic intensity classification, which is used in Japan, and the magnitude of the shaking of an earthquake is interpreted in the system of levels. For example, in 1884, Japan stipulated four levels of seismic intensity, “small earthquake,” “weak earthquake,” “strong earthquake,” and “devastated earthquake.” Currently, the JMA’s Seismic Intensity Level (SIL) is divided into 10 levels, from the seismic intensity level of 0 to the maximum seismic intensity level of 7.

The JMA calculates the seismic intensity observed in each area along with the observed earthquake magnitude and broadcasted the seismic intensity at each point equivalent to a radius of 100 km from the hypocenter as an Earthquake Early Warning on TV. In 1996, the JMA began to use a formula to calculate the seismic intensity level based on Professor Mishio Ishimoto’s research on the shaking observations of surrounding objects, such as houses and trees, during an earthquake (1933).

In Japan, seismic design and construction technology have developed tremendously since 1919. Modern seismic design standards require that buildings not collapse in the event of a major earthquake, even if they undergo major deformation. In contrast to the concept of seismic design standards of buildings based on “strength,” the prevention of “shaking” is still neglected. According to JMA’s records of earthquakes with a seismic intensity of 5 (weak and strong) from 1996 to 2018, the total number of human deaths and those injured was about 98%. The number of collapsed buildings was only about 4%. Hence, most of the causes of human injuries or deaths were not due to the collapse of buildings but due to strong shaking that caused items to fall off shelves, furniture, or ceilings. To solve this problem, besides the concept of “strength” in seismic design codes, it is also necessary to design the quantitative "shaking" of a building.

Challenging toward

Japan is a country of islands in East Asia that presently stands on four active tectonic plates (Pacific plate, North American plate, Eurasian plate, and Philippine plate); thus, Japan is very prone to earthquakes. Japan is also within the Pacific Ring of Fire zone, which is prone to volcanic eruptions. It has been reported since the nineteenth century that 20% of earthquakes worldwide, with a magnitude larger than six, occurred in the vicinity of Japan. By learning from experience, technological developments in Japan for seismic-resistant buildings started in 1891, in the Meiji era. From long experience in developing technologies against the failure and damage of buildings during big earthquakes, the first law enforcement on the seismic-resistant building took place in 1919. Since then, the design codes in Japan have been revised whenever bigger earthquakes and significant casualties have been encountered.

The latest seismic design methods allow the building structures to experience plastic deformation during big earthquakes while remaining elastic under small or moderate earthquakes. The plastic deformation was anticipated to dissipate the earthquake energy and safeguard human beings inside the building by preventing structural collapse. While the current design methods are highly effective for protecting human lives, they do not fully account for human life lost due to non-structural items, such as books, dishes, shelves, furniture, hanging lamps, and ceiling failures due to strong shaking during earthquakes.

There was a continuing tradition where the codes were revised based on the "strength" of the building structures against big earthquakes. A new philosophy in seismic design to reduce shaking effects during earthquakes is necessary to complement the current seismic codes. This new seismic design approach can be applied to more general structures. The such structural design philosophy is required for modern, sustainable, resilient societies. To achieve this, the increase in the construction cost becomes the main issue to resolve; however, the cost of casualties can overwhelm the cost of improving the structural performance. Therefore, this book proposes a quantitative seismic intensity level evaluation during an earthquake to resolve a countermeasure to reduce shaking during earthquakes.

Humans, buildings, and casualties

JMA has comprehensively compiled the statistics of 153 events based on the location, seismic magnitude, Japan Seismic Intensity Level (SIL), and casualties of human and residential housing data from 6th March 1996 until 18th June 2018 (22 years) due to major earthquakes in Japan [2]. Omitting the 2011 Tōhoku earthquake, where the casualties of human beings and residential housing were mostly due to the tsunami, comparisons between the number of human casualties and the collapse of residential housing were investigated.

The earthquake magnitude scales are used to explain the source of energy of an earthquake. These scales differ from SIL values that categorize the intensity or severity of ground shaking caused by an earthquake at a certain ground location. Implicitly, SIL values depend on the magnitude scale, distance and depth of an earthquake hypocenter, type of soil, and dynamic characteristics of the building structure on the ground (see Figure 4.1).

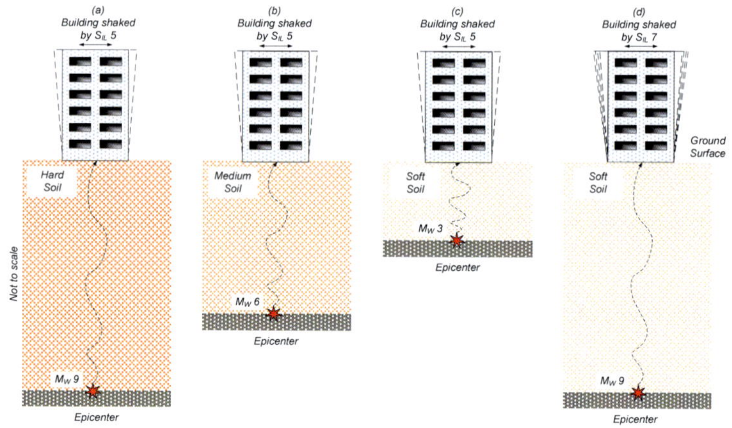

Figure 4.1 The earthquake magnitude and seismic intensity level.

Figure 4.2 shows the statistical data of earthquake magnitude and seismic intensity level at locations from 153 earthquakes recorded. The circles show the magnitude at the hypocenter of the earthquakes and the corresponding Japan Seismic Intensity Level. From Figure 4.2, it can be observed that the magnitude of earthquakes has no relationship with the seismic intensity level.

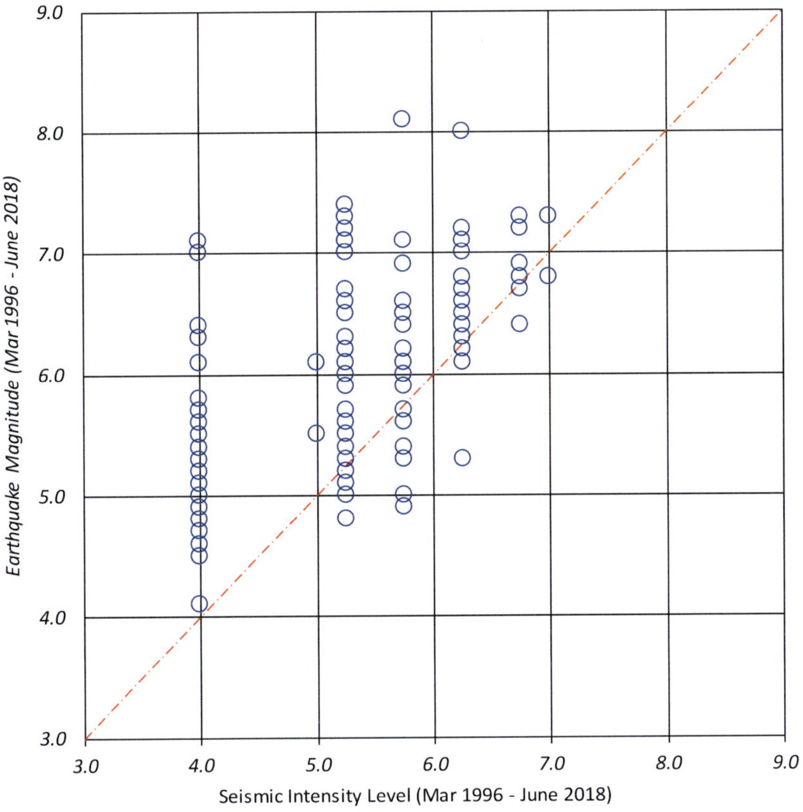

Figure 4.2 The statistical data of the magnitude and seismic intensity level.

Figure 4.3 depicts the statistical data of human casualties (deaths, including missing or injured people) due to earthquakes at different locations. The scatter diamond markers show the number of people injured during the earthquakes. It can be observed that these numbers exceeded the number of deaths, including missing (shown in circles) at the same occurrence of an earthquake.

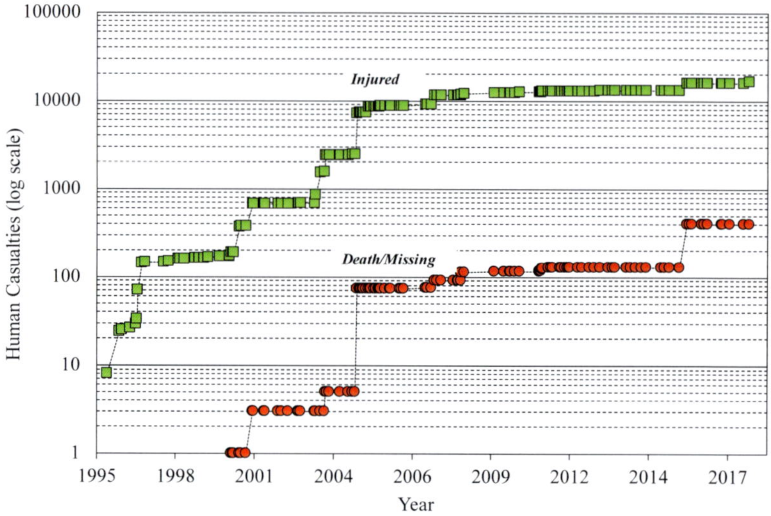

Figure 4.3 The statistical data of human casualties due to earthquakes.

Figure 4.4 depicts the statistical data of residential housing damage (collapsed or repairable and partly damaged) due to earthquakes at different locations. The scatter circle markers show the number of partly damaged residential houses resulting from the earthquakes. It can be observed that these numbers exceeded the number of collapsed residential houses (triangle marker) during the same occurrence of an earthquake.

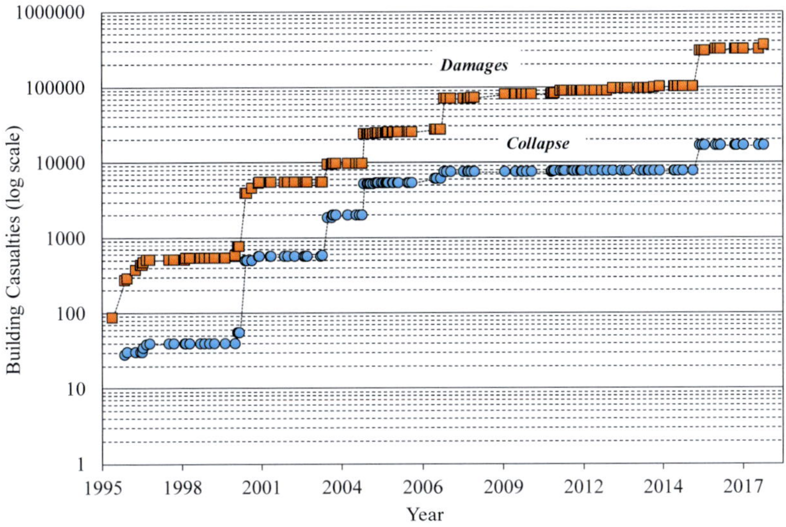

Figure 4.4 The statistical data of building damage due to earthquakes.

Considering all the figures, the number of human beings injured due to earthquakes is greater than the number of deaths/missing. Likewise, the number of collapsed residential houses is less than that of half-collapsed/partly-damaged houses. This implies that residential housing has been proven to have earthquake resistance from several seismic design standard revisions to prevent collapse in a big earthquake. Hence, the human casualties (deaths/injuries) were not mainly due to the collapse of their houses but due to the strong earthquake shaking, which caused items such as books or dishes from shelves, furniture, hanging lamps, and other non-structural elements inside the house to fall, which injured or sometimes killed the people inside the house. In other words, the recent design codes intended to protect human lives against big earthquakes still resulted in casualties due to earthquake-shaking effects.

SIL of Japan

In Japan before 1996, SIL was qualitatively inferred based on the perception of the human body during an earthquake occurrence. The number of people injured due to the earthquake was considered secondary information. Since then, many attempts have been made to quantitatively evaluate the seismic intensity level. JMA has compiled data from about 4,000 seismic stations throughout Japan. In Japan, SIL is widely used to quickly send emergency warnings to the whole country through television and radio broadcasts in the event of an earthquake. This information is to reduce casualties and initiate disaster mitigation.

Table 4.1 describes the 10 levels based on various perceptions of the human body, damage conditions of buildings, and corresponding SIL values (the calculated seismic intensity level is called the *measured SIL*, written as m-SIL).

Table 4.1 Descriptions of SIL and its interpretations.

SIL	m-SIL	Human perception and reaction
0	0–0.4	Imperceptible to people.
1	0.5–1.4	Felt slightly by a few people inside a quiet room of a building.
2	1.5–2.4	Felt by some people inside a quiet room of a building.
3	2.5–3.4	Felt by most people in a building.
4	3.5–4.4	Most people are startled, hanging objects such as lamps swing significantly, and unstable ornaments may fall.
5 Weak	4.5–4.9	Many people are frightened and feel the need to hold onto something stable. Dishes in cupboards and items on bookshelves may fall. Unsecured furniture may move, and unstable furniture may topple over.
5 Strong	5.0–5.4	Many people find it difficult to walk without holding onto something stable. Dishes in cupboards and items on bookshelves are more likely to fall. Unsecured furniture may topple over. Unreinforced concrete-block walls may collapse.

SIL	m-SIL	Human perception and reaction
6 Weak	5.5–5.9	People find it difficult to remain standing. Unsecured furniture moves and topples over. Doors may become stuck shut. Wall tiles and windows may sustain damage and fall. In wooden houses with low earthquake resistance, tiles may fall, and buildings may lean or collapse.
6 Strong	6.0–6.4	People find it difficult to move without crawling. People may be thrown from their position. Most unsecured furniture move and are more likely to topple over. Wooden houses with a low earthquake resistance are more likely to lean or collapse (compared with previous cases). Large cracks may occur, as well as large landslides and collapses.
7	≥ 6.5	Wooden houses with low earthquake resistance are more likely to lean or collapse. Wooden houses with a high earthquake resistance may lean. The majority of reinforced concrete buildings with low earthquake resistance will collapse.

For comparison, the Modified Mercalli Intensity Scale (MMI, [3]), widely adopted in European countries and the United States, is also used to classify shaking effects due to earthquakes. The MMI shows twelve intensity scales with graphical illustrations and brief explanations of the effects on humans of shaking. Still, the perceptions of the human body are essentially a qualitative matter. Some attempts have been made to relate the shaking scale with the magnitude of the hypocenter. However, the relationship between shaking and magnitude neglects the influence of the depth and distance of the hypocenter of earthquakes. On the other hand, the SIL of JMA has ten intensity scales from zero to seven, with refined categories of strong and weak on five and six scales.

Because the SIL values of JMA can be computed based on the input seismic wave and building structure mechanics, it is worth considering it as a more rational approach to calculating the shaking.

SIL denotes how great the strength or force of shaking is due to an earthquake and can be associated with the Peak Ground Acceleration (PGA) and dominant period at a particular ground location. SIL adopted by JMA, shown in Figure 4.5, is very practical because the value of SIL can be quantitatively evaluated from the PGA and dominant period at a given location.

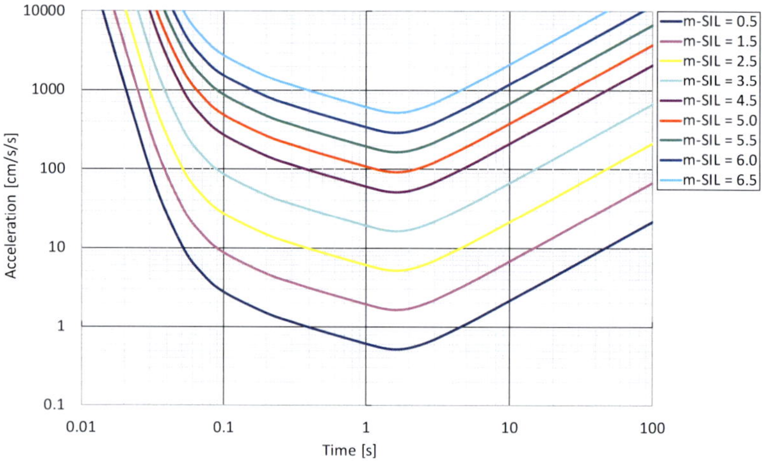

Figure 4.5 Graphs of SIL values.

In practice, SIL values are a necessary evaluation tool in the development of seismic design, seismic diagnosis, seismic strengthening, or seismic strengthening of buildings to reduce the seismic intensity because the quantification of the shaking is more substantial.

An illustrative example of SIL

As an illustrative example, seismic waves of two different earthquakes, as shown in Figure 4.6, which occurred in two locations in Japan within the same year, are selected. First, the Miyagi earthquake occurred on 26th May 2003, with a magnitude of 7.0 at the hypocenter source of the earthquake. In the Ofunato city of Iwate prefecture, the earthquake was recorded to have the Peak Ground Acceleration (PGA) of 1,105.5 gals (cm/s/s) with SIL = 5.8. In the same year, the Tokachi earthquake occurred on 26th September 2003 with a magnitude of 8.0. At that time, in the Urahoro city of Hokkaido, the earthquake was recorded to have a PGA of 454.8 gals with the same SIL = 5.8. Thus, even though both earthquakes were recorded to have different magnitudes and PGAs, they had the same value of SIL = 5.8. The main reasons behind these phenomena were the distance, soil conditions, and depth of the hypocenter of the earthquakes.

Figure 4.6 also shows two different characteristics of seismic waves, the soil condition in Ofunato city has a dominant period of about 0.12 sec (hard soil). In contrast, the soil condition in Urahoro city has a dominant period of about 0.44 sec (soft soil).

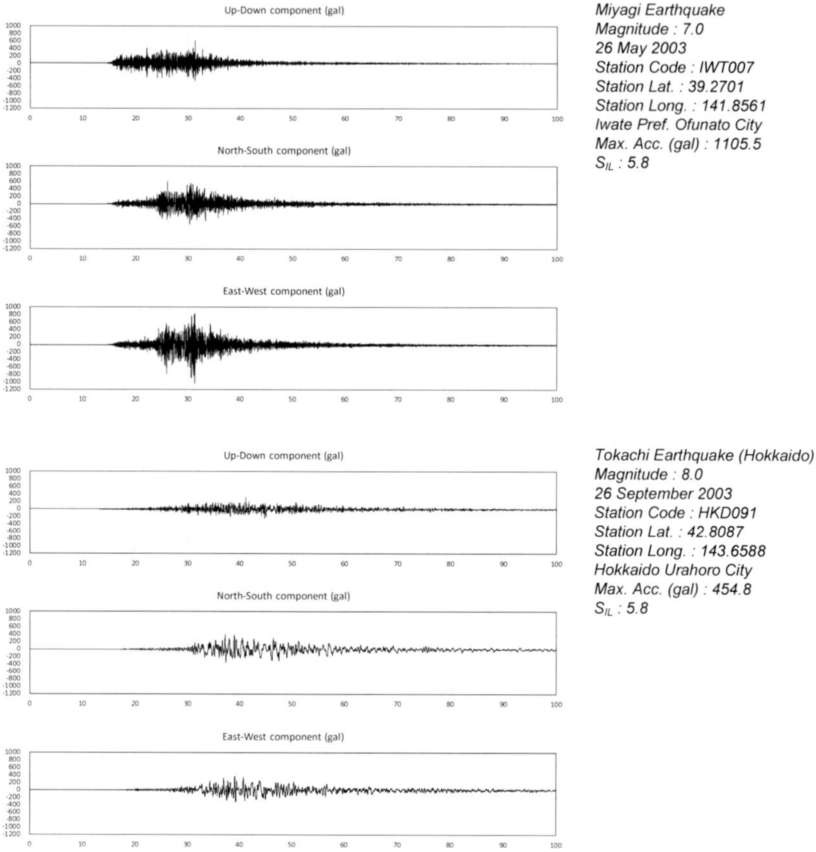

Figure 4.6 Two seismic waves recorded in different locations

When both earthquakes' PGA and dominant period are plotted in Figure 4.7, the same SIL = 5.8 are observed.

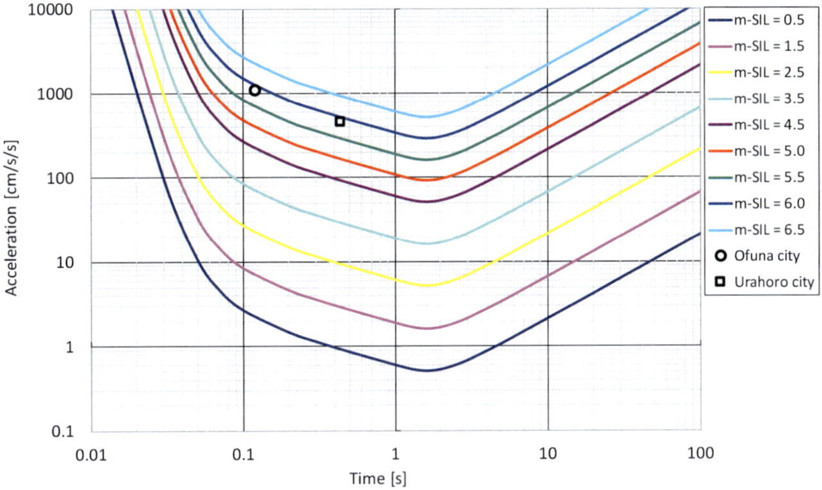

Figure 4.7 Two different cities having the same SIL value

The comparisons imply that the location with large PGA and earthquake magnitude does not have correlations with the seismic intensity SIL or the level of shaking. The relationships between the PGA and SIL or between the earthquake magnitude and SIL are not inherent. However, the shaking level of an earthquake, which can be expressed by the SIL number, is affected by the distance and depth from the hypocenter of the earthquake and PGA. The shaking level SIL can also determine the human and housing casualties during big earthquakes.

Because the SIL of JMA can be evaluated quantitatively, it is an important evaluation tool for determining the efficiency by reducing the seismic intensity at a given location yet being used in the current seismic design method to decrease the number of human casualties even if the houses have not collapsed.

A more comprehensive discussion regarding the proposal to use SIL of JMA as an invaluable criterion for quantitatively evaluating the shaking of buildings in design practices can be found in [4,5].

References

[1] Seismic Intensity Level and Acceleration (in Japanese). *Japan Meteorological Agency.* (http://www.data.jma.go.jp/svd/eqev/data/kyoshin/kaisetsu/comp.htm)

[2] Major Damage Earthquake Occurred in Japan after 1996 (in Japanese). *Japan Meteorological Agency.* (http://www.data.jma.go.jp/svd/eqev/data/higai/higai1996-new.html)

[3] Wood, H.O., and Neumann, F. (1931). Modified Mercalli Intensity Scale of 1931, *Seismological Society of America Bulletin, 21(4)*, 277–283.

[4] Wariyatno, N.G., Han, A. and Gan, B.S. (2019). *Proposed Design Philosophy for Seismic-Resistant Buildings, Civil Engineering Dimension, 21(1)*, 1–5.

[5] Wariyatno, N.G., Han, A., Hsiao, F.P. and Gan, B.S. (2021). *Design Philosophy for Buildings' Comfort-Level Performance, Advances in Technology Innovation, 6(3)*, 157–168.

CHAPTER FIVE

HOW TO CALCULATE SIL

Magnitude and intensity

For the purpose of engineering uses, it is essential to know the size of an earthquake. The most well-known terms, magnitude and intensity are used to express the size of an earthquake. Their definitions are given as follows:

- Magnitude is a measurement of energy released from the hypocenter of the earthquake.
- Intensity is the influence of shaking on a human being, the natural environment, or any building or structure.

The influence of shaking depends on the medium through which the seismic waves propagate, i.e., soil and ground characteristics.

Seismologists measure both quantities by using the scale to describe their sizes.

Magnitude Scales

The impact of an earthquake can be expressed in terms of magnitude. Magnitude is the approximation of energy released, measured as the absolute value of the earthquake without referring to the distance from the earthquake's hypocenter.

The well-known measurement of the earthquake magnitude was introduced by Charles Richter in the 1930s and since then known as the Richter Scale. Richter defined the local magnitude M as the logarithmic of the maximum wave amplitude A in one-tenth of a millimeter recorded on a Wood-Anderson seismograph located 100 km from the hypocenter of an earthquake. The magnitude of an earthquake is defined as

$$M = \log A \tag{5-1}$$

A standard seismograph cannot be set at a distance of 100 km from the hypocenter of an unknown earthquake. In this case, the above formula should be modified as

$$M = \log A - \log A_0 \tag{5-2}$$

where A_0 is the seismographic reading from a standard earthquake given by $A_0 = 0.001$.

Energy is a part of the strain energy released by an earthquake and distributed from the focal region as seismic wave motions. The remainder becomes the potential energy, enabling crustal deformation and energy to be absorbed in the destruction of rocks and slippage at faults. Gutenberg and Richter proposed a correlation between the amount of energy E_f released at the fault and the Richter magnitude given as

$$\log E_f = 4.8 + 1.5M \tag{5-3}$$

where the value of E_f is given in joules. The equation shows that the energy increases in the logarithmic scale of the increase in magnitude M.

Richter's concept behind the local magnitude scale was initially applied to shallow earthquakes and hypocenter distances of less than 600 km. Today the concept has been extended to various magnitude scales on different formulations of the distance from the hypocenter and selecting the suitable wave amplitude.

Surface wave magnitude M_S

Gutenberg defined a magnitude scale based on the measurement of the surface wave, which is recorded in 20-second periods. For an epicentral distance farther away than 2,000 km, the 20-second period is normally dominant on the seismograph records.

Body wave magnitude M_B

There is a traditional method to measure the velocity of the P-wave of deep hypocenter earthquakes. Although the P-wave has a small and insignificant effect like the S-wave, it is faster. It does not depend on the depth of the hypocenter depth.

Seismic moment magnitude M_W

The moment of ground deformation at fault is thought to be more practical and accurate for calculating the magnitude of an earthquake. The moment can be calculated from the product of the dislocation area due to ground deformation and displacement across the fault. The seismic moment magnitude can be approximated by using,

$$M_0 = G\, A_S\, d \tag{5-4}$$

where,

A_S = Area of the dislocation due to deformation of the fault

d = Average displacement across the fault

Hanks and Kanamori [1] proposed a magnitude scale based on the seismic moment as

$$M_W = \log\left(\frac{M_0}{1.5}\right) - 10.7 \tag{5-5}$$

with M_0 in dyne-cm units.

Richter Magnitude M_L

The magnitude of an earthquake is the absolute value of the amount of energy released by the geological rupture, either due to an eruption of a mountain or subduction between two tectonic plates.

The most well-known measurement of earthquake magnitude was introduced by Charles Richter in the 1930s and became famous as the Richter Scale. Richter found that the maximum ground motion decreased logarithmically. The pattern of decreasing is similar for many earthquakes observed.

The magnitude scale of Richter is given by,

$$M_L = \log A(d) - \log A_0(d) \tag{5-6}$$

where,

M_L is the magnitude scale used in the common earthquakes reported by local (shown in subscript L) seismological observatories.

A is the measured maximum trace amplitude of the Wood-Anderson seismograph of the earthquake at the observing station.

A_0 is an empirical function related to a reference earthquake at the observing station.

d is the epicentral distance from the seismic station.

Effects of earthquakes

Table 5.1 describes the typical effects of earthquakes of various magnitude scales near the hypocenter. The magnitude scales are described in size, ranging from micro to great. The magnitude scales are related to their corresponding Mercalli intensity scales. The magnitude scales describe how much the earthquake affects the people and living environments. The table also shows the frequency of occurrences estimated from occurring in seconds to one per 10 to 50 years.

Table 5.1 Typical effects of earthquakes of various magnitudes near the hypocenter [2].

M_L	Size	Mercalli Intensity	Effects	Probability of occurrence
1.0–1.9	Micro	I	Microearthquakes are not felt or rarely felt. Recorded by seismographs.	Million (s) /year
2.0–2.9		I to II	Felt slightly by some people. No damage to buildings.	Million /year
3.0–3.9	Minor	III to IV	Often felt by people but very rarely causes damage. The shaking of indoor objects can be noticeable.	100,000s /year
4.0–4.9	Light	IV to VI	Noticeable shaking of indoor objects and rattling noises. Felt by most people in the affected area. Slightly felt outside. Generally, causes zero to minimal damage. Moderate to significant damage is very unlikely. Some objects may fall off shelves or be knocked over.	10,000-15,000 /year
5.0–5.9	Moderate	VI to VII	It can cause damage of varying severity to poorly constructed buildings. Zero to slight damage to all other buildings. Felt by everyone.	1,000-1,500 /year

M_L	Size	Mercalli Intensity	Effects	Probability of occurrence
6.0–6.9	Strong	VIII to X	Damage to a moderate number of well-built structures in populated areas. Earthquake-resistant structures survive with slight to moderate damage. Poorly designed structures receive moderate to severe damage. Felt in wider areas, up to hundreds of km from the hypocenter. Strong to violent shaking in the epicentral area.	100-150 /year
7.0–7.9	Major		Causes damage to most buildings, some partially or completely collapse or receive severe damage. Well-designed structures are likely to receive damage. Felt across great distances with major damage mostly limited to 250 km from the hypocenter.	10–20 /year
8.0–8.9		X or greater	Major damage to buildings and structures is likely to be destroyed. It will cause moderate to heavy damage to sturdy or earthquake-resistant buildings. Damaging in large areas. Felt in extremely large regions.	One/year
9.0 and greater	Great		At or near-total destruction–severe damage or collapse to all buildings. Heavy damage and shaking extend to distant locations. Permanent changes in ground topography.	One/10–50 years

It should be noted that the values shown are qualitative quantities for explanation purposes because the magnitude scales depend on the characteristics of the soil and ground, distance to the hypocenter, the depth of the hypocenter, and geological conditions (certain terrains can amplify seismic waves).

Intensity scales

An intensity scale is the ground motion intensity scale based on human perceptions against shaking impacts due to an earthquake. The intensity of an earthquake is a measurement of observed damage at a particular location during the occurrence of an earthquake. This intensity will depend on the distance from the hypocenter and local ground conditions. It is necessary to highlight that intensity is a qualitative description of the effects of an earthquake at a particular site. This concept is very suitable for interpreting ancient earthquakes without real recording data. Interpreting ancient earthquake information can help establish the location, recurrence frequencies, and size of the earthquakes in the areas where no seismographs were installed at the time of the earthquake.

The past seismic intensity scales suggested include the Gastaldi Scale (1564) and the Pignafaro Scale (1783). The Rossi–Forel Scale (1883), which has 10 grades, is still used in some parts of Europe. The Mercalli-Cancani-Sieberg Scale, combined with the Mercalli (1902) and Cancani Scales (1904), is still widely used in western Europe. In 1931 Neumann modified the Mercalli-Cancani-Sieberg Scale, proposing a 12-grade Modified Mercalli (MM) Scale, which has now been widely adopted in North America and other parts of the world (Table 5.1). Other intensity scales are the 12-grade Medvedev-Sponheuer-Karnik (MSK) Scale (1964), intended to unify intensity scales internationally, and the 10-grade modified intensity scale of the JMA.

Table 5.2 shows the unabridged (detailed) description of the modified MMI reproduced by Stover and Coffman [3] from the original abridged (outline) version by Wood and Newmann [4]. The Rossi–Forel (RF) scale of intensity (values are ranging from I to X) was used in the late nineteenth century. The RF scale was replaced by the Modified Mercalli Intensity (MMI) scale in 1931 [4]. The MMI scale has twelve grades denoted by Roman numerals from I to XII, describing the physical effects of earthquakes.

Table 5.2 Unabridged Modified Mercalli Intensity (MMI) Scale.

Intensity value	Description
I	Not felt except by a very few under especially favorable conditions.
II	Felt only by a few persons at rest, especially on upper floors of buildings.
III	Felt quite noticeably by persons indoors, especially on upper floors of buildings. Many people do not recognize it as an earthquake. Standing motor cars may rock slightly. Vibrations are similar to the passing of a truck. Duration estimated.
IV	Felt indoors by many, outdoors by few during the day. At night, some awakened. Dishes, windows, doors disturbed; walls make cracking sound. Sensation like heavy truck striking building. Standing motor cars rocked noticeably.
V	Felt by nearly everyone; many awakened. Some dishes and windows are broken. Unstable objects overturned. Pendulum clocks may stop.
VI	Felt by all. Many frightened. Some heavy furniture moved: a few instances of fallen plaster. Damage slight.
VII	Damage is negligible in buildings of good design and construction; slight to moderate in well-built ordinary structures; considerable damage in poorly built or badly designed structures; some chimneys are broken.
VIII	Damage is slight in specially designed structures; considerable damage in ordinary substantial buildings with partial collapse. Damage is great in poorly built structures. Fall of chimneys, factory stacks, columns, monuments, walls. Heavy furniture overturned.
IX	Damage is considerable in specially designed structures; well-designed frames are thrown out of plumb. Damage is great in substantial buildings, with partial collapse. Buildings shifted off foundations.
X	Some well-built wooden structures are destroyed; most masonry and frame structures are destroyed with foundations. Rails bent.
XI	Few, if any (masonry) structures remain standing. Bridges destroyed. Rails bent greatly.
XII	Damage total. Lines of sight and level are distorted. Objects are thrown into the air.

Table 5.2 shows that the intensity scales are established based on visible damage and human perceptions. Therefore, there is no information regarding the maximum acceleration of ground motion, and a correlation with the magnitude is missing. The seismic intensities depend on the location of the hypocenter, the direction of the seismic wave from the focus to the observation station, geological conditions below the observation points, and other influences.

Countries like Japan and Indonesia, prone to strong earthquakes, have published leaflets of graphical information to disseminate the rate of shaking during the occurrence of earthquakes for public viewing. Figures 5.1 and 5.2 represent seismic intensity levels in Indonesia and Japan, respectively.

Figure 5.1 Illustration of Modified Mercalli Intensity Scale by BMKG (Indonesia, https://www.bmkg.go.id/gempabumi/skala-mmi.bmkg?lang=EN).

Figure 5.2 Illustration of Seismic Intensity Scale by JMA (Japan, https://www.jma.go.jp/jma/en/Activities/inttable.html#files).

The intensity of an earthquake is a measurement of the observed effects and damage level of a building at a particular location. The intensity will vary according to the distance from the epicenter, the depth of the hypocenter, and soil and ground condition characteristics. It should be noted that intensity is a qualitative description of the effects of an earthquake. This concept is very useful in interpreting historical observed earthquake data.

The intensity level of JMA can be calculated and categorized into level groups, as shown in Table 5.3. The calculation of m-SIL values will be discussed in the following section.

Table 5.3 Descriptions of the SIL (freely translated from Japanese).

SIL	m-SIL	Human perception and reaction
0	0–0.4	Imperceptible to people.
1	0.5–1.4	Felt slightly by a few people inside a quiet room of a building.
2	1.5–2.4	Felt by some people inside a quiet room of a building.
3	2.5–3.4	Felt by most people in a building.
4	3.5–4.4	Most people are startled, hanging objects such as lamps swing significantly, and unstable ornaments may fall.
5 Weak	4.5–4.9	Many people are frightened and feel the need to hold onto something stable. Dishes in cupboards and items on bookshelves may fall. Unsecured furniture may move, and unstable furniture may topple over.
5 Strong	5.0–5.4	Many people find it difficult to walk without holding onto something stable. Dishes in cupboards and items on bookshelves are more likely to fall. Unsecured furniture may topple over. Unreinforced concrete-block walls may collapse.
6 Weak	5.5–5.9	People find it difficult to remain standing. Unsecured furniture moves and topples over. Doors may become wedged shut. Wall tiles and windows may sustain damage and fall. In wooden houses with low earthquake resistance, tiles may fall, and buildings may lean or collapse.
6 Strong	6.0–6.4	People find it difficult to move without crawling. People may be thrown from their position. Most unsecured furniture moves and is more likely to topple over. Wooden houses with a low earthquake resistance are more likely to lean or collapse (compared with previous cases). Large cracks may occur, as well as large landslides and collapses.
7	≥ 6.5	Wooden houses with low earthquake resistance are more likely to lean or collapse. Wooden houses with a high earthquake resistance may lean. The majority of reinforced concrete buildings with low earthquake resistance will collapse.

Calculation of JMA SIL

The procedure for computing the m-SIL value of the JMA intensity scale was proposed by Kawasumi in 1943 [5]. A brief explanation is presented herein for clarity.

Figure 5.3 describes the procedure for computing the m-SIL value of JMA proposed by Kawasumi [5]. It should be noted that the accelerations used in calculating m-SIL are not the maximum raw value of the ground acceleration. However, they must be transformed into the frequency domain and filtered and sorted. The acceleration used in calculating the m-SIL value can be found after the resultant accelerations are sorted at 0.3s of time.

The meaning of the symbols in Figure 5.3 is given below.

- $a_{x,y,z}(t) = a_x(t), a_y(t), a_z(t)$: time domain response acceleration of a certain location in x, y, z directions, respectively.
- $A_{x,y,z}(t) = A_x(t), A_y(t), A_z(t)$: frequency domain response acceleration of a certain location in x, y, z directions, respectively.
- $A'_{x,y,z}(t)$: filtered $A_{x,y,z}(t)$ by human sensitivity to shaking.
- $A''_{x,y,z}(t)$: filtered $A'_{x,y,z}(t)$ by a high-cut (low-pass) band.
- $A'''_{x,y,z}(t)$: filtered $A''_{x,y,z}(t)$ by a low-cut (high-pass) band.
- $a'_{x,y,z}(t) = a'_x(t), a'_y(t), a'_z(t)$: inversed time domain response acceleration of a certain location in x, y, z directions.
- $a_{0.3}$: the minimum response acceleration defined at 0.3 seconds continuously sorted from the maximum response acceleration.

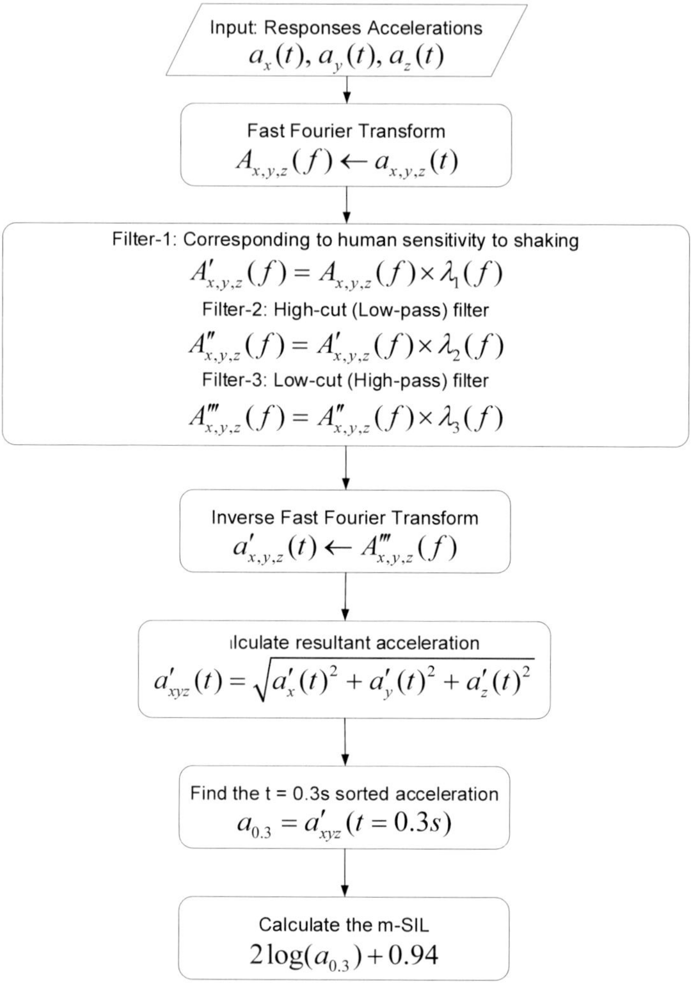

Figure 5.3 Flowchart to calculate SIL value m-SIL.

What is Fourier Transform?

As shown in Figure 5.3, the Fast Fourier Transform applies to each of the three-component acceleration records into the frequency domain of the ground responses. And the Inverse Fast Fourier Transform is used to transform back the acceleration from the frequency domain to the time domain.

Most signals, such as seismic waves, are time domain signals which measure acceleration obtained from the seismograms/accelerometer devices. That is, the seismic wave signal in terms of acceleration is a function of time. In other words, when we plot a signal, the horizontal axis is the time, and the vertical axis is usually the amplitude of accelerations. The most important information is often hidden in the signal's frequency domain. That hidden information in the frequency domain is used to calculate the m-SIL of JMA.

How to determine $a_{0.3}$ acceleration

Figure 5.4 illustrates how to determine the $a_{0.3}$ values. Figure 5.4(a) shows the bandpass-filtered absolute resultant acceleration response in the time domain. The dotted line defines the lowest maximum acceleration $a_{0.3}$ of a total continuous 0.3 seconds around the peak acceleration. Figure 5.4(b) shows an easy method to determine acceleration $a_{0.3}$ by sorting the resultant accelerations in Figure 5.4(a) and taking the resultant acceleration value at the cumulative duration of 0.3 seconds.

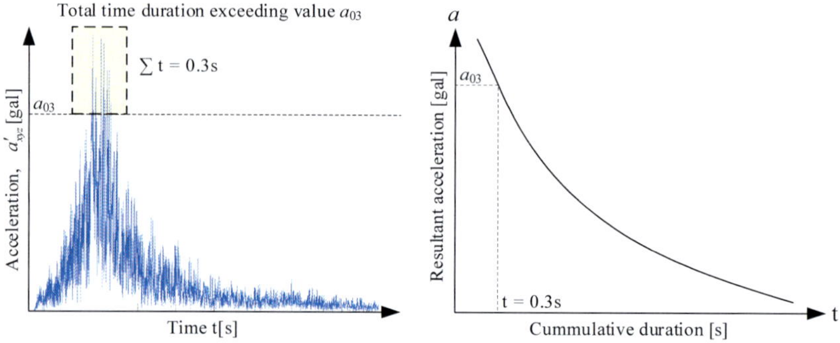

Figure 5.4 Total time duration 0.3s to determine the value of $a_{0.3}$.

Frequency band of Kawasumi filters

Three frequency bandpass filters of Kawasumi are given in the following equations.

$$\lambda = \lambda_1 \times \lambda_2 \times \lambda_3 \tag{5-7}$$

$$\lambda_1 = \left(\frac{1}{f} \right)^{0.5} \tag{5-8}$$

$$\lambda_2 = \begin{pmatrix} 1 + 0.694 y^2 + 0.241 y^4 + 0.0557 y^6 + \\ 0.009664 y^8 + 0.00134 y^{10} + 0.000155 y^{12} \end{pmatrix}^{-0.5} \tag{5-9}$$

$$\lambda_3 = \left(1 - e^{-8 f^3} \right)^{0.5} \tag{5-10}$$

$$y = \frac{f}{10} \tag{5-11}$$

where f represents the dominant frequency, λ_1 the filter on-period effect, the high-cut filter, and the low-cut filter. Figure 5.5 shows the filter characteristics of the seismic intensity.

These filters are applied to the frequency domain of the three-component accelerations (North-South, East-West, and Up-Down directions) of seismic waves from an earthquake. Before applying the filters, the seismic waves of an earthquake recorded in monitoring stations must be transformed from the time domain to frequency domain signals. For a discrete signal, the discrete Fast Fourier Transform method is used to transform the domain of the signals. Moreover, after the filters are applied, the three directions of acceleration in the frequency domain are then transformed back to the time domain. The normalized vectored resultants of the three components are used to calculate the $a_{0.3}$ acceleration.

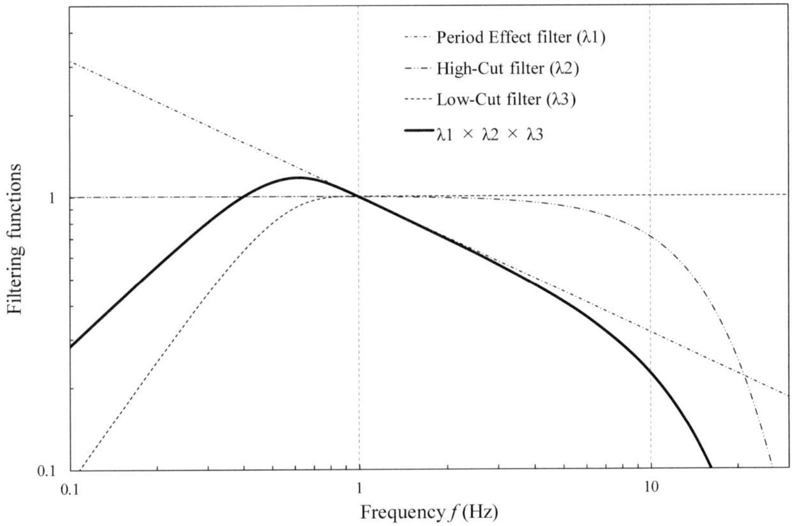

Figure 5.5 Three frequency bandpass filters of Kawasumi [5].

Example: 1940 El Centro Earthquake

The 1940 El Centro seismic waves illustrate the process of determining the m-SIL value of the earthquake. The earthquake occurred at 21:35 Pacific Standard Time on May 18 in the Imperial Valley in south-eastern Southern California near the international border of the United States and Mexico. It had a seismic moment magnitude M_W of 6.9 and a maximum perceived intensity of *X* (Extreme) on the Mercalli intensity scale. It was the first major earthquake recorded by a strong motion seismograph located next to a hypocenter. The earthquake was characterized as a typical moderate-sized destructive event. It caused widespread damage to irrigation systems and led to the deaths of nine people. Figure 5.6 shows the recorded El Centro seismic wave components.

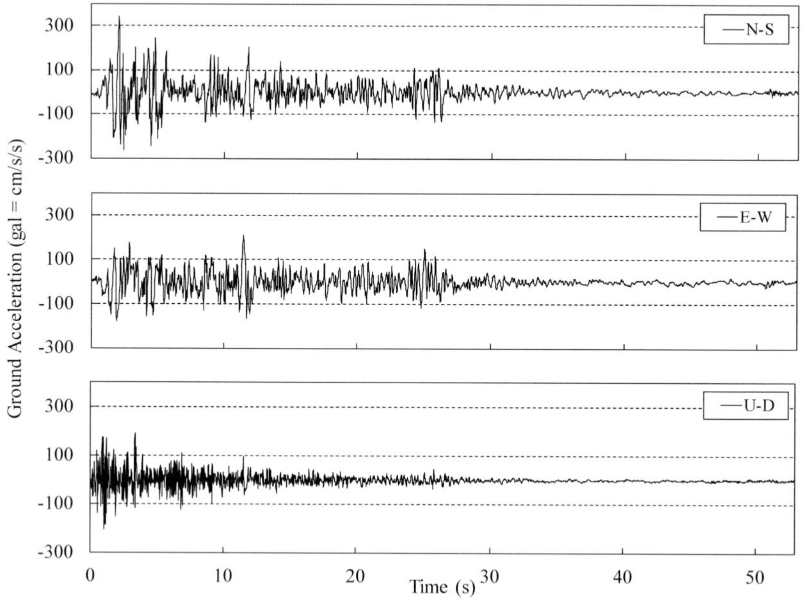

Figure 5.6 El Centro seismic waves (N-S, E-W, and U-D) in the time domain.

Figure 5.7 shows the recorded El Centro seismic wave components after transforming from the time domain to the frequency domain using the Discrete Fast Fourier Transform method.

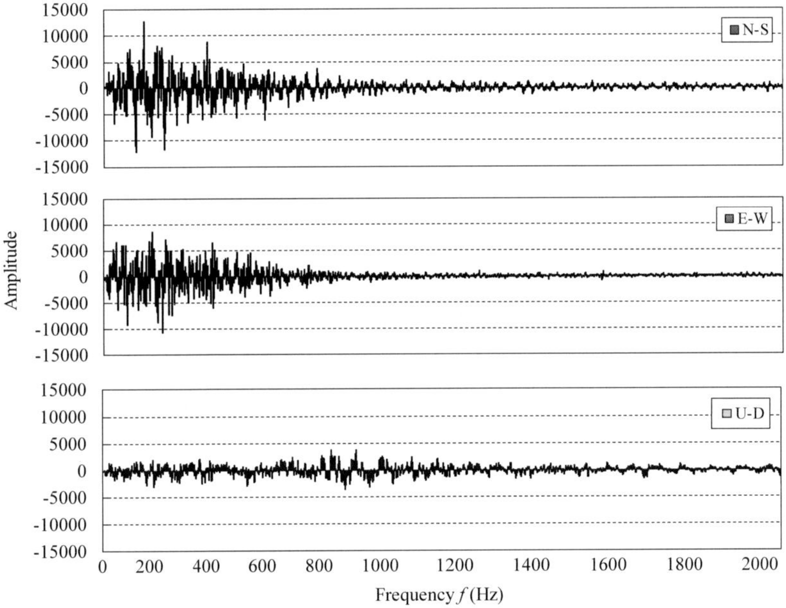

Figure 5.7 El Centro seismic waves (N-S, E-W, and U-D) in the frequency domain.

Figure 5.8 shows the recorded El Centro seismic wave components after applying three frequency bandpass filters of Kawasumi [5].

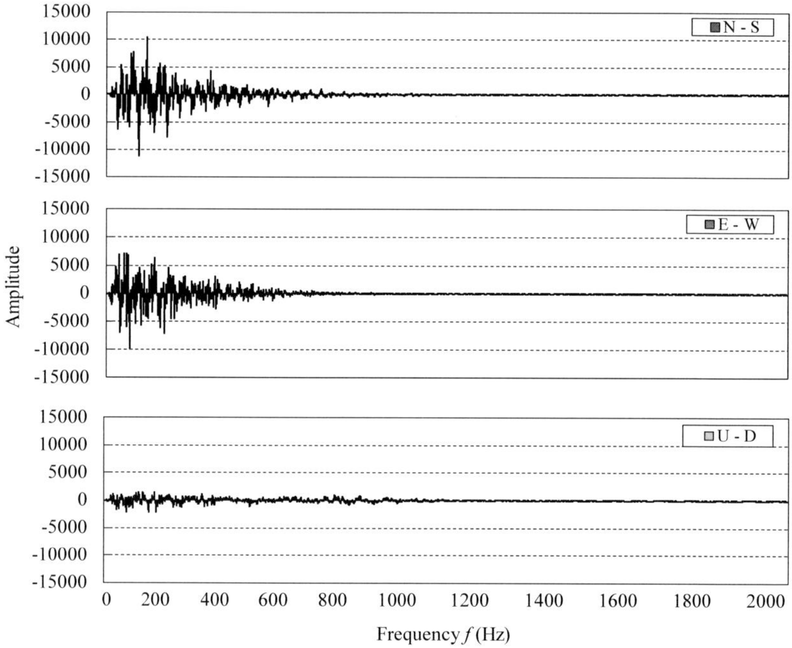

Figure 5.8 After applying three frequency bandpass filters of Kawasumi [5].

Figure 5.9 shows the recorded El Centro seismic wave components after transformation from the frequency domain to the time domain using the Inverse Discrete Fast Fourier Transform method.

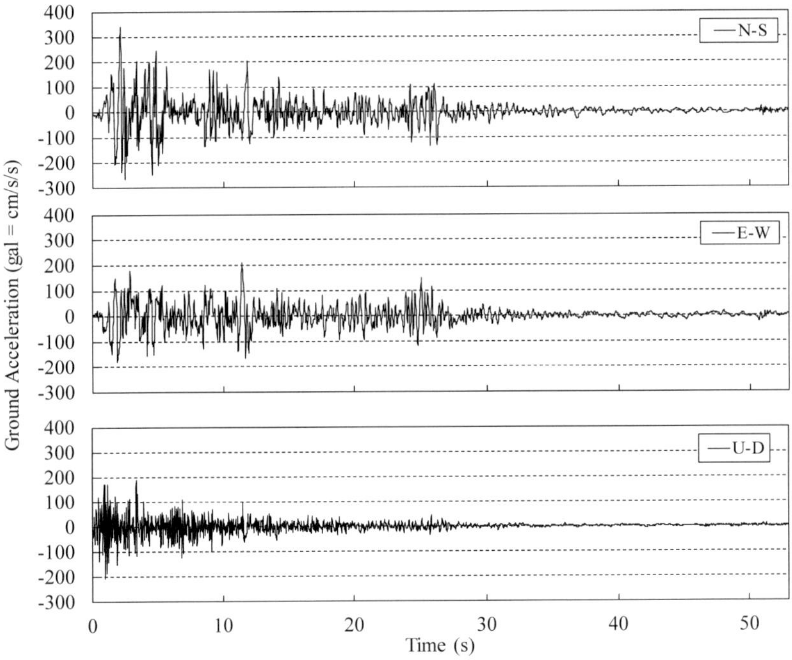

Figure 5.9 After inversing from frequency to time domain by using IFFT.

Figure 5.10 shows the recorded El Centro seismic waves' resultant acceleration after vectorizing three components of accelerations.

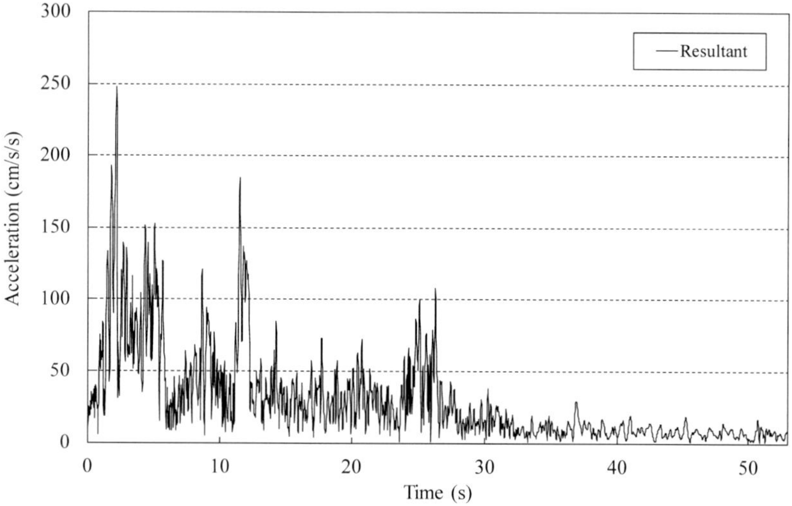

Figure 5.10 Resultant acceleration in the time domain.

Figure 5.11 shows the recorded El Centro seismic waves' resultant acceleration after being sorted in descending order. From where the value of $a_{0.3}$ can be determined.

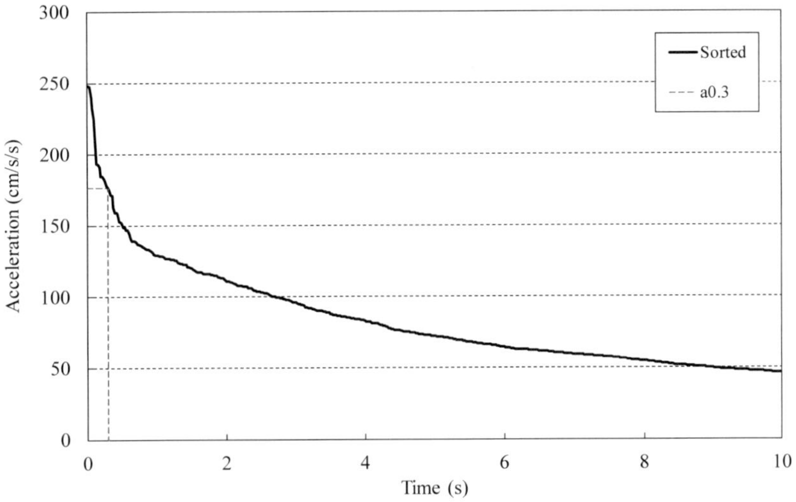

Figure 5.11 Resultant of accelerations shown for the first 4 seconds.

By using the filtered time domain acceleration and its resultant vectored acceleration, the value of m-SIL can be determined using the following:

$$m - SIL \text{ value} = 2\log\left(a_{0.3}\right) + 0.94 \quad (5\text{-}12)$$

where $a_{0.3}$ is the acceleration of a continuous total duration of 0.3 seconds from the maximum acceleration response. From Figure 5.11, the m-SIL value of the 1940 El Centro earthquake at the recording station is found to be:

$$a_{0.3} = 176.64\ cm/s^2$$
$$m - SIL \text{ value of El Centro} = 5.43 \tag{5-13}$$

FFT can determine the dominant period of El Centro on the ground. The result of the seismic wave in the N-S direction of the El Centro earthquake is shown in Figure 5.12.

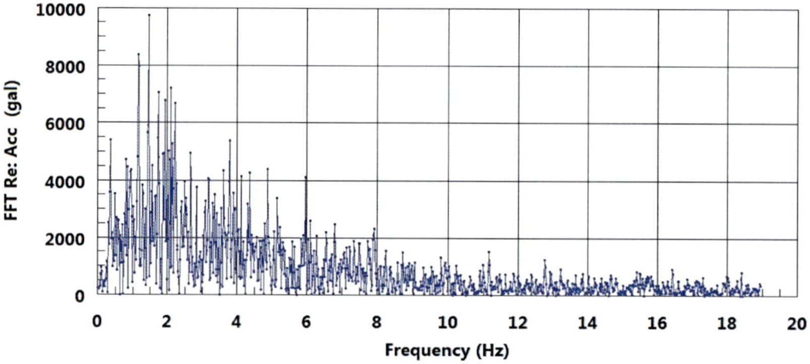

Figure 5.12 The flow to determine the dominant period of the seismic waves.

From Figure 5.12, the dominant period of ground acceleration of El Centro N-S direction waves is located at the peak of FFT real numbers on the frequency:

$$f_0 = 1.46279 \text{ Hz}$$
$$t_0 = 1/f_0 = 0.6836 \text{ sec} \tag{5-14}$$

The following flowchart, as shown in Figure 5.13, uses Equations (5-7) to (5-11) to calculate the value of $A_{0.3}$.

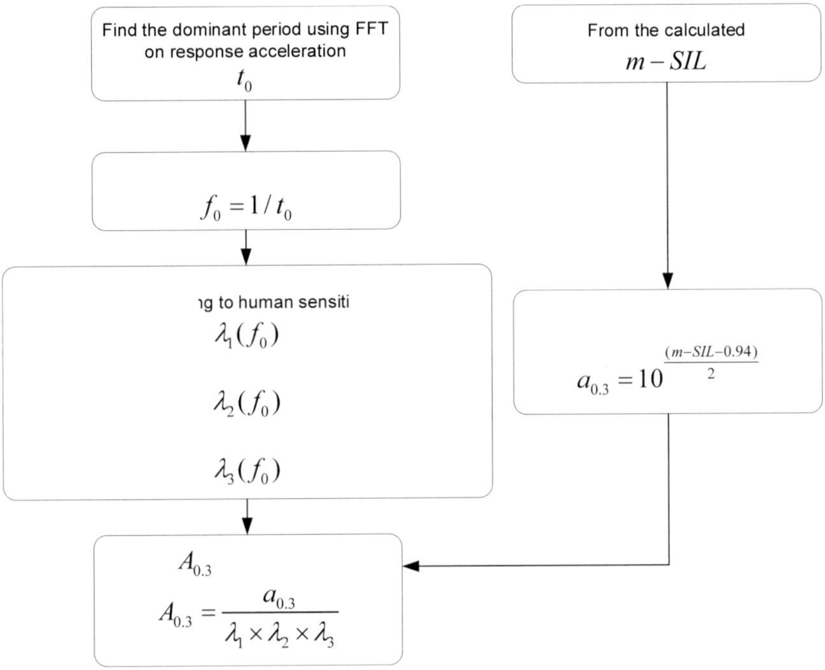

Figure 5.13 The flow to determine $A_{0.3}$ of the seismic waves or response waves.

The calculated $A_{0.3}$ results in:

$$A_{0.3} = 214.2 \text{ gals} \tag{5-15}$$

The value of $A_{0.3}$ with respect to the time $T = t_0$ is plotted in Figure 5.14. Figure 5.14 shows the m-SIL value of the 1940 El Centro earthquake to be 5.2, close to the 5.5 Curve, the 5 Strong intensity level of SIL.

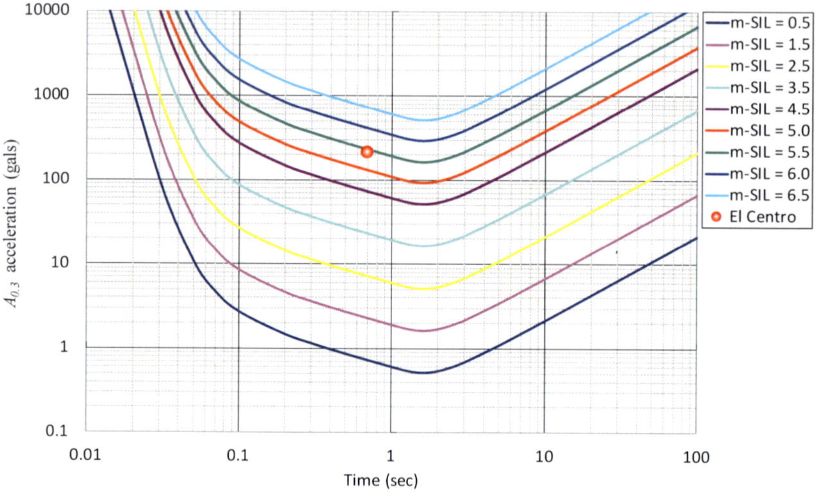

Figure 5.14 The m-SIL value of the 1940 El Centro Earthquake.

To show graphically how strong the shaking was during the 1940 El Centro earthquake. JMA's illustrative description, as shown in Figure 5.15, can be used as a physical explanation of the shaking quantity. Therefore, the benefit of quantification of shaking by using the JMA SIL value is recommended for practical use.

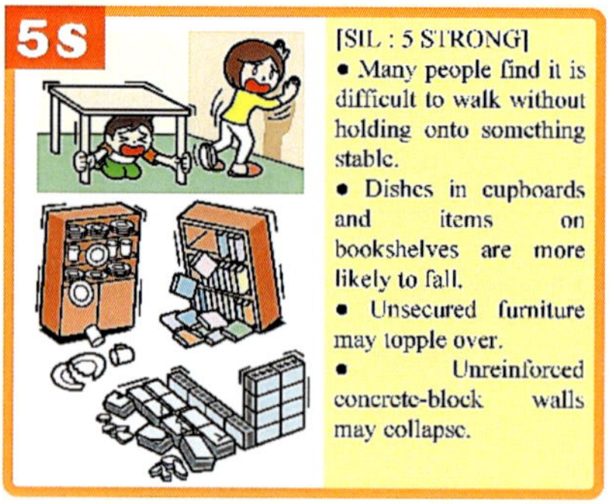

Figure 5.15 JMA Illustration of shaking quantity 5 Strong of SIL.

References

[1] Hanks, T.C. and Kanamori, H. (1979). A moment magnitude scale, J Geophys Res, 84:2348–2350.

[2] United States Geological Survey. "Earthquake Facts and Statistics." (https://web.archive.org/web/20100524161817/http://earthquake.usgs.gov/earthquakes/eqarchives/year/eqstats.php) Retrieved May 01, 2022.

[3] Stover, C.W., and Coffman, J.L. (1993). Seismicity of the United States, 1568-1989 (Revised): U. S. Geological Survey Professional Paper 1527, 418.

[4] Wood, H.O., and Neumann, F. (1931). Modified Mercalli Intensity Scale of 1931, *Seismological Society of America Bulletin, 21(4)*, 277–283.

[5] Kawasumi, H. (1943). Measures of Earthquake Danger and Expectancy of Maximum Intensity Throughout Japan as Inferred from the Seismic Activity in Historical Times, Earthquake Research Institute, Vol. XXIX, No. 43, 469–482.

CHAPTER SIX

HOW TO MEASURE THE SHAKING OF EARTHQUAKES

What is a seismic wave?

Earthquakes produce seismic waves when one body of solid material, like soil, slides against another; the interface between the two bodies is the fault. Suppose the equilibrium of a solid body like the soil is disturbed due to earthquakes triggered by the fault's movement. In that case, the seismic waves will be transmitted through the soil layers in all directions from the earthquake's focus. Earthquakes can radiate seismic waves within periods of tenths of seconds to several minutes. Rocks behave like elastic solids at these frequencies. Elastic solids allow various wave types, making measuring ground motion after an earthquake difficult.

There are two types of seismic waves: body waves and surface waves, which travel from the focus of an earthquake.

Body waves

The body wave propagates in an infinite continuum in the same direction as its vibration. Primary (P)-wave (longitudinal or compressive wave). The body wave propagates in an infinite continuum in the perpendicular direction of its vibration is called a Secondary (S)-wave (transversal or shear wave). P- and S-waves are transmitted from the inside of the ground from the focus to the surface.

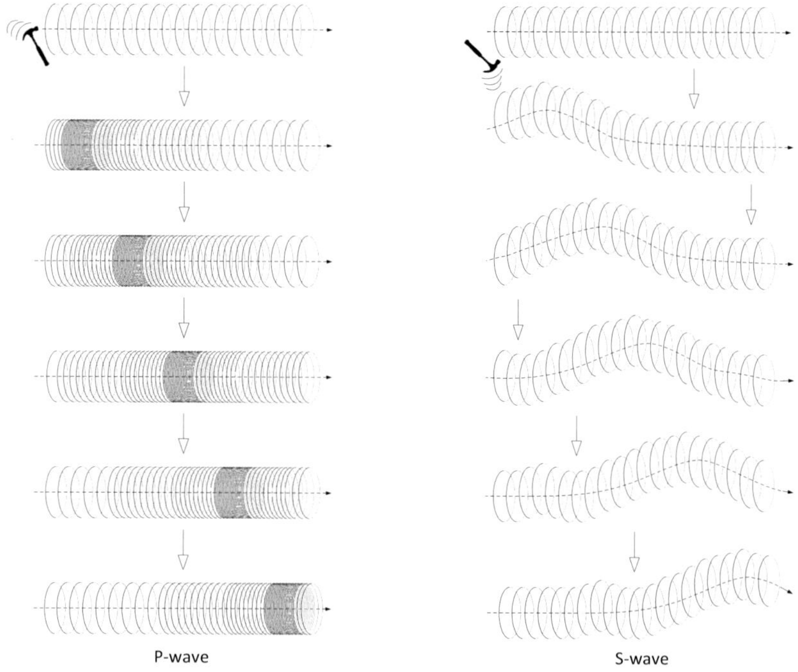

Figure 6.1 Illustrations of body waves (P-wave and S-wave).

The velocities of the P- and S-waves depend upon the elastic constants and densities of the material and other properties of the surrounding medium, soil.

The propagation velocities of the P- and S-waves V_P and V_S are expressed by

$$V_P = \sqrt{\frac{E}{\rho}\frac{(1-\nu)}{(1+\nu)(1-2\nu)}} \tag{6-1}$$

and

$$V_S = \sqrt{\frac{G}{\rho}} = \sqrt{\frac{E}{\rho}\frac{1}{2(1+\nu)}} \tag{6-2}$$

where,

E = Young's elastic modulus

G = Shear modulus

ρ = Mass density

ν = Poisson ratio

Generally, in any soil material, the relationships $V_P > V_S$ held. Suppose the Poisson ratio of the ground is taken to be $\nu = 0.25$, then by using Equations (6-1) and (6-2), we can obtain a relationship as,

$$V_P = \sqrt{3}\, V_S \tag{6-3}$$

Equation 6-3 shows that the P-wave will arrive at an observation station faster than the S-wave because its velocity is higher. The first waves to arrive are the P ("Pressure" or compression) waves (Figure 6.1 left). P-wave has its roots in the Latin word *primus*, which means first, since they are the first waves to arrive, having the highest velocity. The P-wave transmits to an observation station earlier than the S-wave. In the earthquake accelerograms, the P-wave is recorded for a time before the S-wave arrives.

The P or compression waves transmit pressure changes through the Earth in alternating compressions and rarefactions. Since P-waves arrive earlier than S-waves, P-waves are most commonly used to find an earthquake's hypocenter and other related fields in seismic exploration.

The S-waves arrive following the P-waves because the waves have lower velocities. The second waves to arrive are the S ("*Secundus*") waves which have a transverse, shear vibration in a plane perpendicular to the direction of propagation (Figure 6.1 right). Because the S-waves are perpendicular to their propagation direction, especially on the ground surface, the waves can cause damage by shaking due to shearing forces.

Surface waves

Surface waves propagate on the ground surface. In shallow earthquakes, surface waves are frequently detected. Surface waves produce large amplitude motions on the ground surface and disappear much more slowly than body waves, resulting in maximum damage. Surface waves are more damaging because of their low frequency, long duration, and large amplitude. The types of surface waves are L-waves (Love waves) and R-waves (Rayleigh waves). The L-waves vibrate longitudinally and perpendicularly to the ground surface and display an elliptic movement. L-waves' velocity is less than or close to that of S-waves (Figure 6.2). An S_v-wave is one in which the ground motion (vibration) is upright, and an S_h-wave refers to one where the ground motion is horizontal (on the plane).

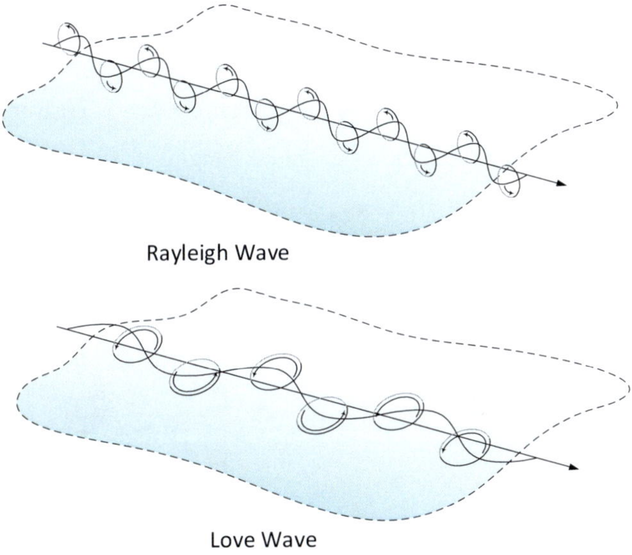

Figure 6.2 Illustrations of surface waves (R-wave and L-wave).

Rayleigh waves, also known as "ground-roll," are the effect of interaction between P- and S-waves and are similar to oceanic waves. The presence of these waves was first revealed by an English Physicist, John William Strutt (Lord Rayleigh), in 1885. Rayleigh waves may be noticeable in wide open fields during an earthquake. Love waves are the effect of the interaction between P- and S_h-waves. The presence of these waves was first revealed by a British Mathematician A.E.H. Love, in 1911. The Love waves transmit faster than Rayleigh waves.

All about seismometers

Seismograph

Seismic waves are usually measured and recorded by using a seismograph. A seismograph can assist seismologists, geologists, and scientists in measuring the magnitude and location of the hypocenter of an earthquake. The principle of the seismograph is that ground motion is measured from the vibration record of a simple pendulum hanging from a stable frame (Figure 6.3).

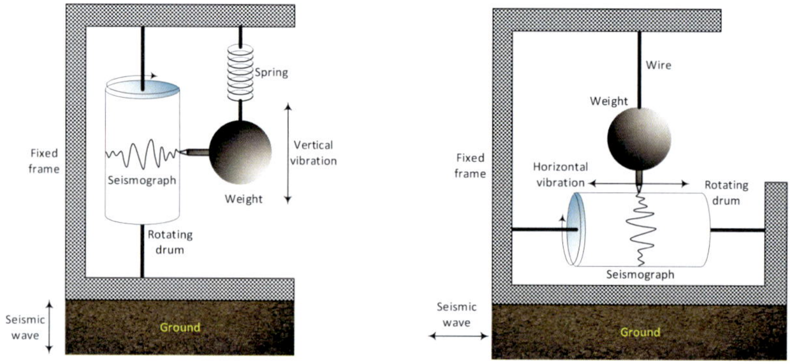

Figure 6.3 Principle of a seismograph.

In the system shown in Figure 6.3, by choosing an appropriate damping coefficient and setting the natural period of the pendulum as relatively long compared to the period of the ground motion, the pendulum's velocity becomes proportional to the ground motion v_g . Therefore, the recorded velocity v can be correlated with the velocity v_g of the ground motion. This kind of seismograph is known as a velocity seismograph or a long-period seismograph.

The pendulum's damping coefficient can be obtained by setting the period of the pendulum to be relatively short compared with the period of the ground motion when the acceleration a approaches the gravity acceleration a_g. This implies that the ground acceleration due to the ground motion can also be recorded using a seismograph. This kind of seismograph is known as an acceleration seismograph or a short-period seismograph. Setting the damping coefficient of the pendulum to a large value and letting the natural period of the pendulum be close to that of ground motion results in $a \propto a_g$. This implies that the ground velocity due to the ground motion can also be recorded by using a seismograph.

Figure 6.4 shows a long-period seismograph. The recording would not be possible without making the pendulum instrument long enough in the system. A simple pendulum and an inverted pendulum can measure a long-period pendulum.

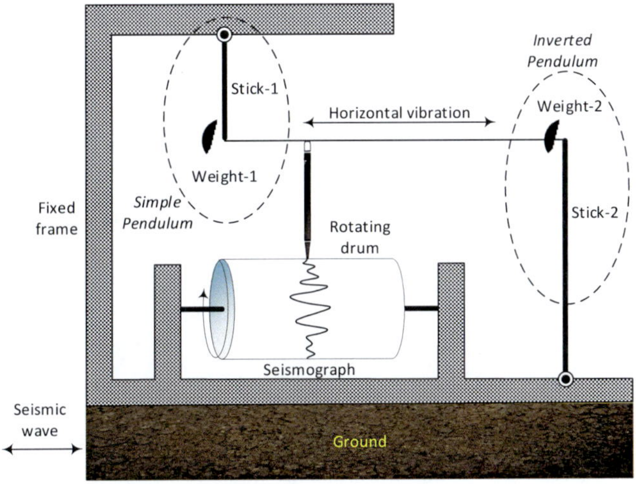

Figure 6.4 Folded pendulums for long-period seismograph [1].

John Milne invented the modern seismograph in 1880 while at the Imperial College of Engineering in Japan. In the late nineteenth century, most seismographs were developed by a team of John Milne, T.A. Ewing, and others working in Japan (1880–1895). Most seismographs are manufactured based on their damped inertial-pendulum systems. By permitting the seismic waves to move the frame, the suspended mass stays in a vertical position so that the seismometer measures the relative motion between the frame and the mass.

Strong motion accelerograph

It was realized that the displacement seismograph, based on pendulum systems, had many limitations, especially during a strong earthquake. Although descriptions of damage due to earthquakes have been reported over the last 2,000 years, most of the data collected were qualitative assessments of the disasters. There was almost no progress in the measurement of seismic activity until strong motion-recording instruments were developed.

The next developed instrument was a teleseismometer. This is an instrument that can record a broad range of frequencies of seismic waves. A servo feedback system keeps the mass motionless, and the relative movement of the frame is measured by an electronic displacement transducer with minimum force applied to the mass.

Another type of seismograph that can record digital strong motion data is an accelerograph. An accelerograph can measure accelerations, and the recorded data can be integrated to yield the velocity and displacement of a strong-motion earthquake. The data recorded from an accelerograph is important for understanding how an earthquake affects the structure of a building. An accelerograph is not as sensitive to ground motion as a teleseismometer. Still, even the strongest earthquake hits are stable on their scales. The measured accelerations are three seismic waves that correspond to the vertical [U-D], north-south [N-S], and east-west [E-W] accelerations. The instrument was aligned southward, so the transverse horizontal acceleration appeared on the east-west component. The first detected seismic wave arrival is the P-wave that produces ground motion along the direction of wave propagation. The L-wave occurs on the transverse motions of the E-W component, and the R-wave occurs only on the vertical and north-south components.

Approximately 2,200 strong motion accelerographs were installed worldwide up to 1974, and the number is said to have doubled shortly after that (Cherry, 1974). Borcherdt and Matthiesen (1980) show details of a strong motion instrumentation program in the United States. The records of a strong motion accelerograph may include errors caused by factors such as:

(1) frequency features of the accelerograph

(2) elongation and twisting of the recording paper

(3) offset, deviation, and friction of the recording pen

(4) errors in reading the record. These errors are usually corrected by filtering and other techniques

Field observation

To provide a reliable basis for the seismic design of building structures, earthquake occurrences should be assessed as accurately as possible. In recent decades, numerous efforts have been made to collect records of many strong earthquakes in many countries. The characteristics and size of an earthquake depend not only on the earthquake's source mechanism but also on geological site conditions, soil site conditions, and other influencing factors. Hence the record of an after-quake disaster taken in one place sometimes becomes the standard for understanding the earthquake's mechanism.

1989 Loma Prieta post-earthquake observation

On October 17, 1989, an earthquake with a magnitude of 6.9 severely struck the San Francisco and Monterey Bay areas. The hypocenter was located at approximately 37.0 degrees North Latitude and 121.9 degrees West Longitude, near the Loma Prieta Peak in the Santa Cruz Mountains of San Francisco (Figure 6.5).

Contemporary design methodologies and construction practices have shown inadequacies against stronger earthquakes, resulting in structural failures, lifeline facilities discontinuation, and loss of human life. Previous lessons have improved engineering analysis, design, and construction practices.

(a) Landslides on the San Mateo Coast

(b) Liquefaction in the Watsonville area

I Building damage in San Francisco

(d) Ground failure in Watsonville Area

(e) Collapsed highway in Oakland

(f) Collapsed highway in Watsonville Area

Figure 6.5 Field observations after the 1989 Loma Prieta Earthquake [2].

Seismic tomography

Seismic tomography (seismic imaging) is a technique for imaging the ground using the seismic waves of an earthquake. Body waves (P-waves and S-waves) and surface waves can produce three-dimensional models of the ground interior.

Tomography is solved as an inverse problem. The data received at seismometers determine the location of reflection and refraction of the seismic waves' path. To accurately predict the earthquake's hypocenter

Figure 6.6 illustrates the paths of seismic waves from the hypocenter through the soil layers of the ground to the seismometers in stations on the surface.

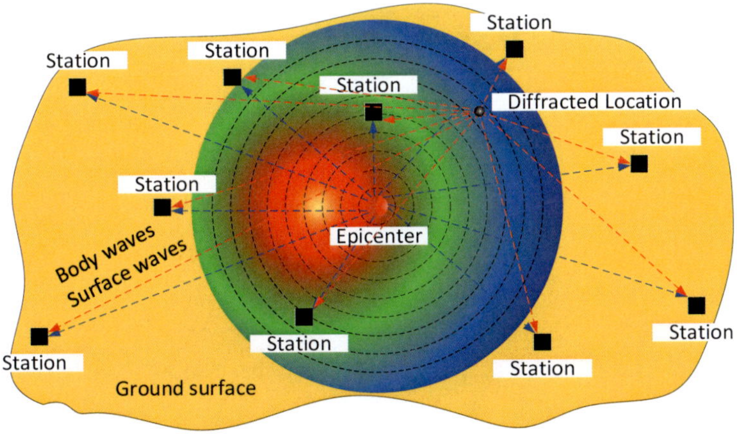

Figure 6.6 Illustration of seismic wave paths in aerial view.

Seismic records of ground motions (commonly, acceleration) as a function of time provide the basic data seismologists use to study elastic waves spreading throughout the ground. Seismology is an observation-based science that tries to interpret seismograms. There is also the physical constraint of recording seismic wave motions only at (or very near) the ground's surface. Hence seismologists greatly benefit from mathematical concepts to solve equations called inverse theory.

Modern accelerometer

Accelerometers have many uses in science, the military, and industry. Highly sensitive accelerometers are used in inertial navigation systems for aircraft and missiles. Moreover, vibration in rotation machines has to be monitored by accelerometers. Accelerometers are also widely used in tablet computers, smartphones, and digital cameras, so images on the screens are always adjusted vertically. In uncrewed aerial vehicles such as drones, accelerometers help to stabilize their flight in the air.

Electronic accelerometer

Since the invention of the seismograph, people have been relegated to bulky, expensive conventional accelerometers. This challenged wide-ranging research and the development of smaller and more compact electronic accelerometers to replace the mechanical-based accelerometer.

There are two types of electronic accelerometers, AC- and DC-activated devices (Figure 6.7). AC accelerometer devices have an AC-coupled output of electrical signals, and DC accelerometer devices have a DC-coupled output of electrical signals. The main difference is that, with a DC response, the output of electrical signals can drive down to zero Hz (Hertz), making it easier to sense small changes in motion. DC-based accelerometers are more commonly used than AC-based accelerometers since they are not equipped with the same level of sensitivity to measure smaller changes in vibration.

Figure 6.7 AC (left) and DC (right) accelerometers (image used courtesy of TE Connectivity).

These accelerometers can be found in large engineering applications such as measuring vibration on cars, machines, buildings, process control systems, and safety installations. They can also measure seismic activity, inclination, machine vibration, and dynamic speed.

In physics, the acceleration principle relates to the inertia of a moving body converted to a force unit. Applying this rule, an object's acceleration can be converted into current, charge, or voltage as the outputs.

MEMs

Meanwhile, a small footprint for electronic accelerometers is required to advance high computer performance and smaller devices' computing capability. This led to extensive research and development of MEMs (Micro-Electro-Mechanical system) technology. We can see a seamless integration of MEMs accelerometers into portable commercial devices such as smartphones, shock monitoring equipment, and vibration measurement devices used in various engineering practices. The use of MEMs technology in accelerometers has helped reduce size and costs. The appeal of MEMs accelerometers also lies in their ability to operate in low temperatures, reduce noise levels and offer low power dissipation. However, they suffer from having poor signal-to-noise ratio and limited dynamic range.

Figure 6.8 shows 3-axis gyroscopes and 3-axis accelerometers suitable for applications where a high-performance angular motion sensor is required. This MEMs accelerometer supports a wide range of engineering applications.

Figure 6.8 Three-axis MEMs Accelerometer (Image used courtesy of TDK).

Seismic intensity measuring device

Traditional accelerometer measuring systems, which play important roles in earthquake engineering, tend to be wired network systems and are impractical. A portable seismic intensity measuring device is necessary for the context of practical and essential seismic wave measurements.

Raspberry Pi

Raspberry Pi is a low-cost (less than 100 USD) credit-card-sized computer that can be plugged into a display or monitor and uses a standard keyboard and mouse (Figure 6.9). Raspberry Pi, developed by the Raspberry Pi Foundation and associated with Broadcom, is a product line of small single-board computers and perhaps the most exciting small-size computer available today. The Raspberry Pi will act as a small personal computer by connecting peripherals like a keyboard, mouse, display, and extension boards.

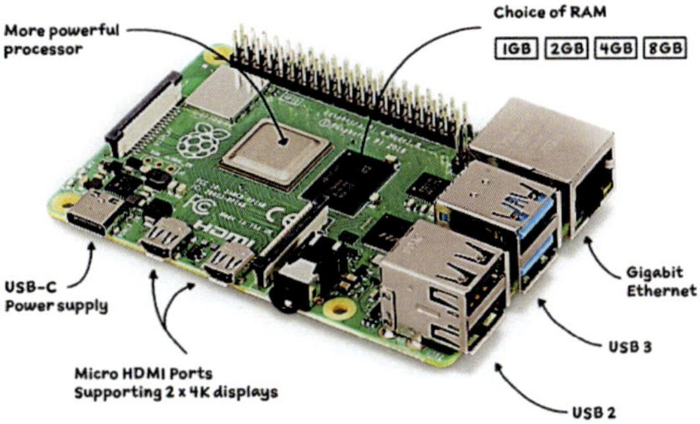

Figure 6.9 Raspberry Pi 4 model (www.raspberrypi.com).

Raspberry Pi is a fully customizable and configurable, integrated solution for reliable and accurate seismic intensity identification at a particular location during an earthquake. Although Raspberry Pi is slower than a laptop or desktop, it is still a computer that can provide all the expected features or abilities with low power consumption. Thus, it is suitable as a portable monitoring device.

The CPU clock speed of the Raspberry Pi used is 1.8 GHz, and it has an onboard SDRAM of 4 GB. An adapter board should be plugged into the main board to connect the extension boards of the accelerometer and display functionality from the main board. As for the adapter board, the Pmod HAT adapter developed by Diligent connects the main board with extension boards (Figure 6.10).

Figure 6.10 Pmod HAT adapter plugged into the Raspberry Pi.

The extension board of the accelerometer used in this study is PmodACL2, with a physical size of 2.41 cm by 2.03cm (Figure 6.11). The Digilent PmodACL2 is a 3-axis MEMS accelerometer that uses the ADXL362 chip to run up to twelve bits of resolution for each axis of acceleration. Additionally, this module offers a variety of feedback from multiple interrupt sources and power-saving features through its inactivity monitoring.

Figure 6.11 Three-axis MEMs Accelerometer inside PmodACL2 board.

To be used as a handy small-size personal computer, we need a display to show the results of measurements of seismic waves. The 2.13-inch e-Paper HAT produced by Waveshare is used as a peripheral connected to the main board by plugging it into the Pmod HAT adapter (Figure 6.12).

Figure 6.12 2.13-inch e-Paper HAT extension board (3.02 cm by 6.5 cm).

After plugging the extension boards of the accelerometer and displaying components to the main controller board of Raspberry Pi (Figure 6.13), we are ready to use this small-size personal computer to measure seismic waves and compute the seismic intensity to be shown on display.

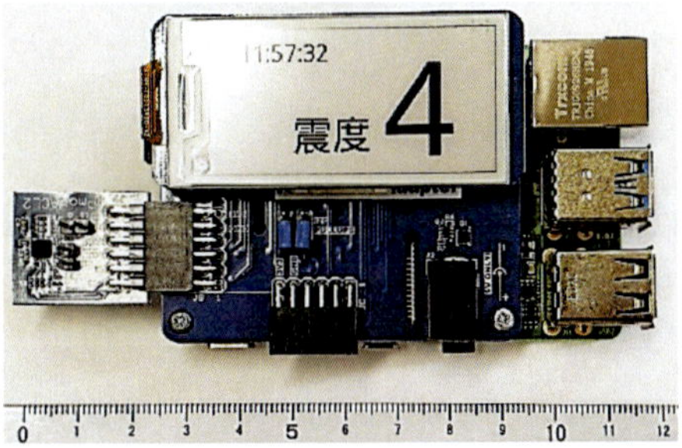

Figure 6.13 A handy small-size seismic wave monitoring system.

Testing the portable device

Currently, SIL is being used in Japan to disseminate an early warning notification to people in the surrounding area of the occurrence of strong earthquakes. The SIL values broadcast to the public are the intensity of shakings on the ground surface at particular locations. With the shaking of floors in a moderately high building, the SIL values are usually larger than at the ground surface due to the mechanical behavior of the building. Therefore, developing a portable device that can detect the shaking of each floor in the building is needed. An instrument to measure seismic waves and, at the same time, calculate the SIL value has many practical applications. To test the customized portable device, the shaking table shown in Figure 6.14 was used as the bedrock to produce artificial seismic waves.

Figure 6.14 Shaking table in Nihon University (Sanesu Co. Ltd.).

The 5-story building was modeled using a simple aluminum frame structure (see Figure 6.15). The importance of the test was not to imitate the entire building but to verify the customized SIL device in measuring the SIL value correctly. Six acceleration transducers (ARH-20A) were installed on each plate level in the frame. The ARH-A is a series of acceleration transducers from the Tokyo Measuring Instruments Laboratory company. The ARH-20A acceleration transducer has a 20 m/s^2 (~2G) measurement capacity and operates at 150 Hz with only an 18-gram weight.

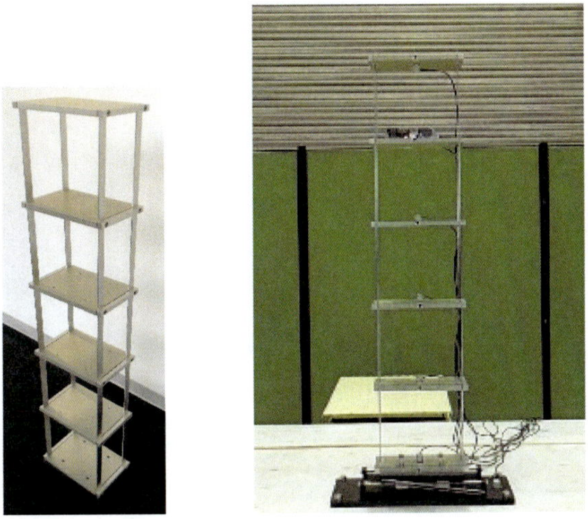

Figure 6.15 Installation of portable SIL device (on the fifth floor).

During the experiment, the shaking table was given a sine wave with varying natural frequencies as the input wave. When the rotating frequencies of the input wave and the frame are close, the frame will show a resonance condition where it shakes with a large deflection. Figure 6.16 shows the accelerations on the frame's fifth floor measured using the ARH-20A acceleration transducer and portable device. Following the calculation procedure described in Chapter Five, the test results in the SIL had a value of 5.3 for both devices.

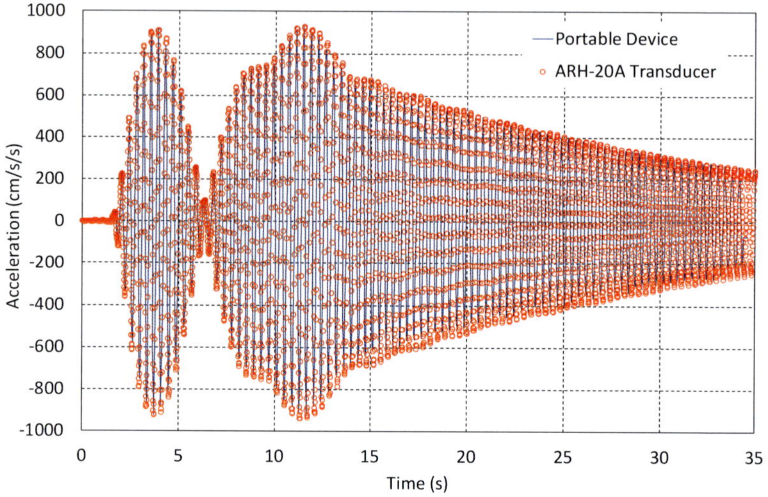

Figure 6.16 Measurements of accelerations on the fifth floor of the frame.

From Figure 6.16, the measured accelerations recorded by the customized portable device were in good agreement with the ARH-20A. Thus, the built portable device can measure seismic waves and simultaneously calculate the SIL values on every building floor.

References

[1] Bertolini, A., Riccardo, D., Francesco, F. and Akiteru, T. (2006), Monolithic Folded Pendulum Accelerometers for Seismic Monitoring and Active Isolation Systems, IEEE Trans. Geoscience and Remote Sensing 44, 271–276.

[2] Nakata, J.K., Meyer, C.E., Wishire, H.G., Tinsley, J.C., Updegrove, W.S. Peterson, D.M., Ellen, S.D., Haugerud, R.A., McLaughlin, R.J., Fisher, G.R., and Diggles, M.F., The October 17, 1989, Loma Prieta, California Earthquake – Selected Photographs, U.S. Geological Survey, Digital Data Series DDS-29 (URL:https://pubs.usgs.gov/dds/dds-29/, retrieved on 20 April 2022).

CHAPTER SEVEN

SIL AS AN EVALUATION TOOL

Applications of SILs for building

SIL's many potential applications include an evaluation tool for a seismic building. Instead of the "strength" concept of current evaluation tools, a SIL comes with a quantitative "comfort" evaluation tool that provides physical meanings on how an earthquake's strong shaking can impact human lives in a building. The following are general concepts that illustrate the application of SILs to a building.

In this chapter, several test cases were conducted to demonstrate the real applications of SILs in practice. The test cases of buildings' existing, strengthening, and vibration absorption systems were considered in the study. The shaking performance evaluations of strengthening methods and installation of vibration control mechanisms in the existing building were compared using SILs values and economic considerations. The trade-off concept was used in this situational decision that involved losing control to shaking in return for gain in the economic aspect.

Existing building prototype

As an example of an existing building prototype, a 5-story steel frame building was considered. The building plan was 12 m by 15 m, and the total height was 18.1 m. The building was built following the Japanese seismic codes [1]. Figure 7.1 shows the existing building prototype (steel structure) demonstrating the performance evaluations of strengthening methods and vibration absorption systems.

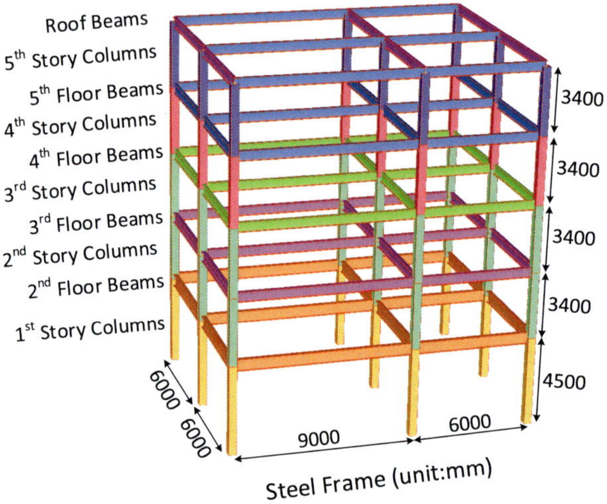

Figure 7.1 A 5-Story Steel Frame Building.

The steel structure comprised wide-flange beams and square box column sections (Table 7.1). All the beams and columns had fully restrained connections.

Table 7.1 Cross sections of beams and columns.

Floor	Story	Beam	Column
Roof		W-500x200x9x16	
	5		□-400x400x12
5		W-500x200x9x19	
	4		□-400x400x16
4		W-500x200x9x22	
	3		□-400x400x19
3		W-500x250x9x22	
	2		□-400x400x19
2		W-600x250x12x22	
	1		□-400x400x22

Input ground motion

In this study, the recorded 1940 El Centro seismic wave of the North-South (N-S) component was used as the ground motion input for the time history dynamic analysis of the building prototype (Figure 7.2).

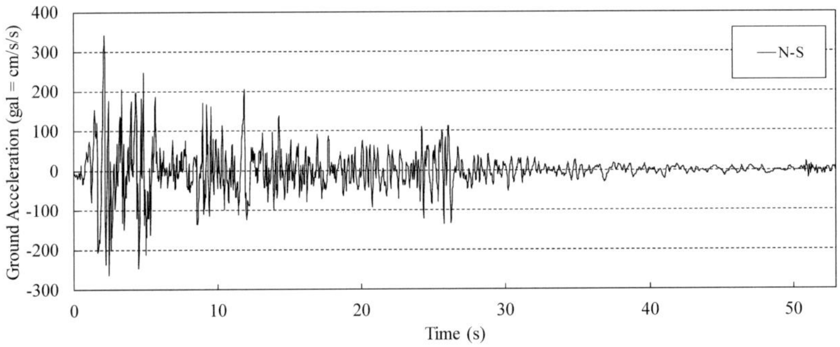

Figure 7.2 El Centro seismic waves (N-S) in the time domain.

The N-S component had a Peak Ground Acceleration (PGA) of 342 gals (cm/s^2).

Planning and design of a building

In the planning and design phases of a building, the designer has an important role in analyzing the proposed drawings prepared by the architectural designer. A designer is responsible for designing a building structure that exceeds the owner's expectations while ensuring the resistance and strength of the structure. A designer is also responsible for performing a structural design (Figure 7.3) that can function safely over the building's useful life. The perception of the useful life of a building implies considerations of durability. It generates the basis for considering the growing degradation of time-varying risks (e.g., seismic loads, occupant loads, wind loads, etc.).

Given that performance and cost are inseparably linked, the designer must consider economic limitations to the main goals of safety and durability. As discussed in Chapter Four, reducing human casualties from earthquakes involves not only increasing the resistance and strength of the building. Even though the building is designed to meet the latest seismic design provisions, there are uncertainties regarding the shaking due to strong earthquakes. In other words, the building would undoubtedly shake by an undeterminable amount when a strong earthquake strikes the location. Therefore, it is very attractive if, in the planning and design phases, the shaking level during strong earthquakes can be designed for our building after the completion of construction.

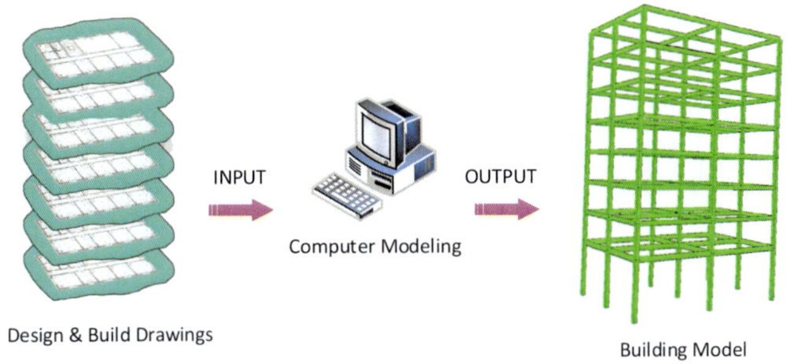

Figure 7.3 Concept of planning and structural design of a building.

With the new philosophy in seismic design, the current seismic design practice has added the concept of "comfort" by designing the "shaking" from the beginning of the planning phase. Although adding the "comfort" value to a building will increase the cost of the building, a suitable solution could be achieved by conducting a feasibility study which is more reliable because the shaking behavior of the building (Figure 7.4) can be predicted quantitatively.

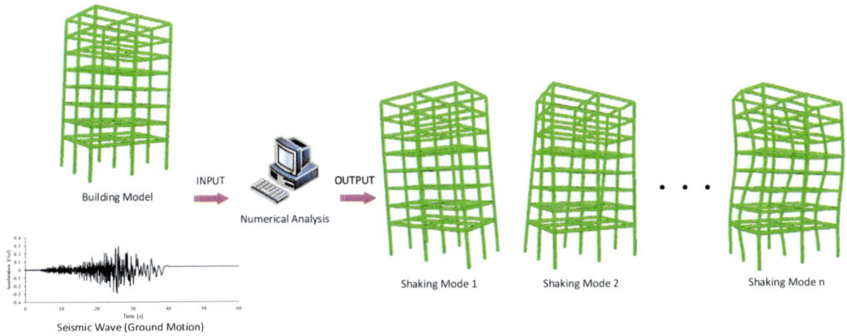

Figure 7.4 Prediction of building's shaking behavior in the design phase.

By identification of SIL values on each floor in a building, we can estimate how strongly the floor will shake during an earthquake, as required by the seismic design codes (Figure 7.5). Thus, we can reduce the shaking for the "comfort" of occupants in the building. Reducing the shaking will require a higher cost to be a trade-off.

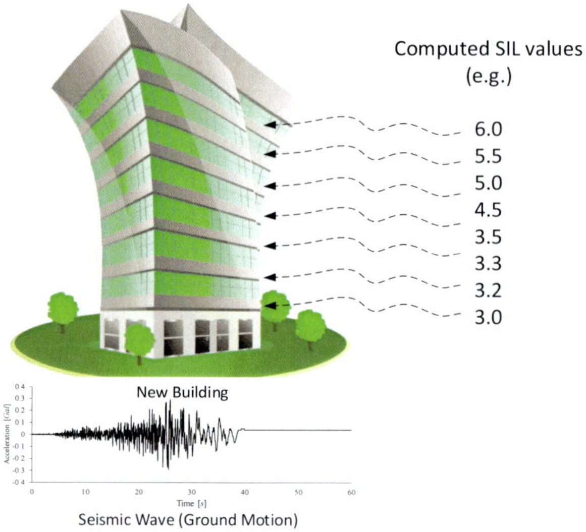

Figure 7.5 Identification of SIL values on each floor of a building in the design phase.

Shaking performance evaluation of existing building

The shaking performance evaluation of an example building prototype with the 1940 El Centro as the input ground motion is shown in Table 7.2 and Figure 7.6. It can be seen in Table 7.2 that the $A_{0.3}$ accelerations and calculated m-SIL values increase in the height direction.

Table 7.2 Results of m-SIL calculations of the existing building.

Floor	m-SIL	T (sec) [FFT]	$A_{0.3}$ (gals)	Natural Frequencies of Building Prototype
Roof	6.8	0.2643	1739.9	
5	6.7	0.2643	1550.7	Mode 1: T = 0.2621 sec
4	6.5	0.2643	1231.8	Mode 2: T = 0.2173 sec Mode 3: T = 0.1870 sec
3	6.2	0.2643	872.0	Mode 4: T = 0.0942 sec
2	5.8	0.2643	550.2	Etc.
Ground	5.2	0.6836	164.4	

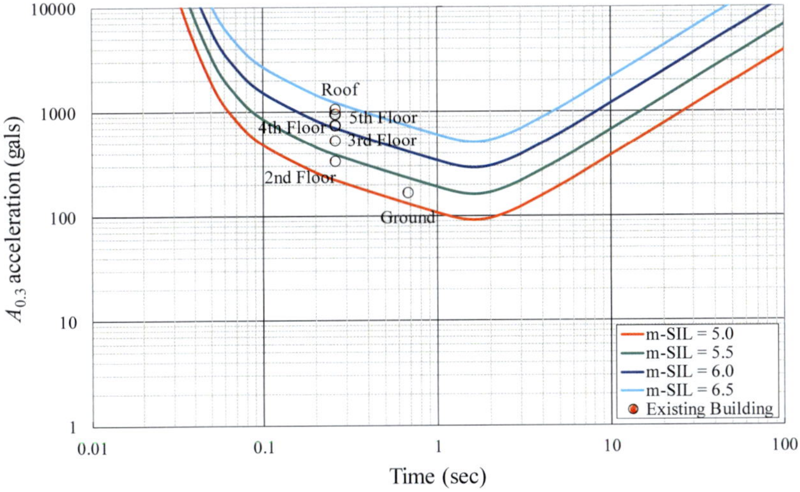

Figure 7.6 SIL of a building prototype in planning and design.

As demonstrated above, we can determine the m-SIL values on each floor in a building. From the calculated m-SIL values, the maximum accelerations at each position can be computed following the procedure described in Chapter Five. By plotting the values of the dominant period and a_{max} accelerations in the SIL graphs, we can see that the m-SIL at the ground surface is much less than on the roof (Figure 7.6). In applying the present methodology, Figure 7.6 should be used with Figure 7.7 as accompanying illustrative shakings of each floor in the building.

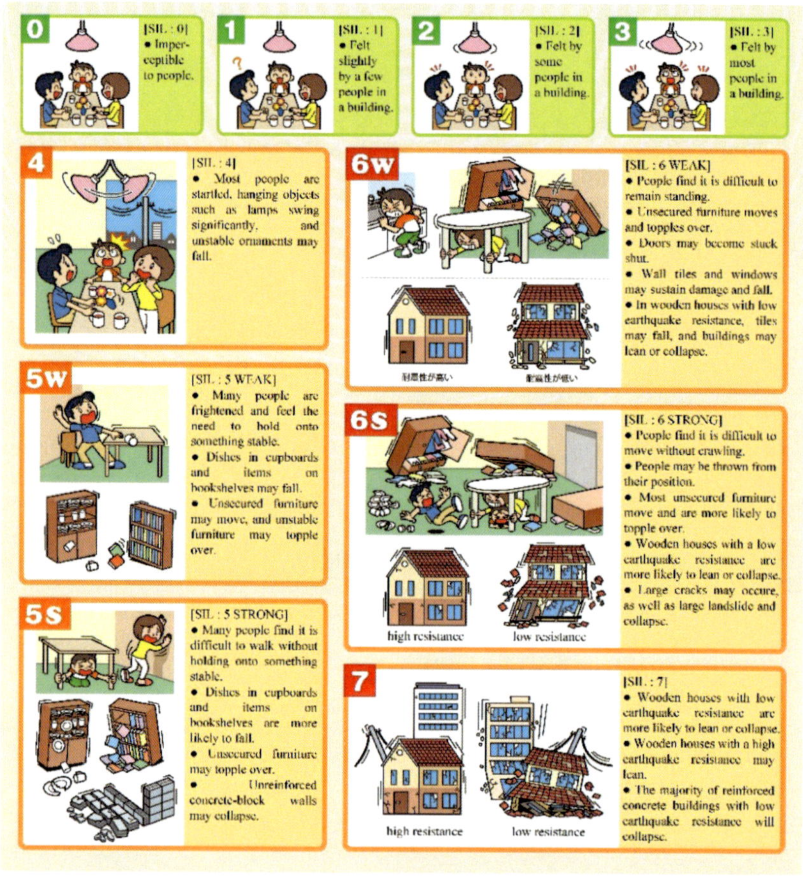

Figure 7.7 Illustrative quantification of JMA'S SILs (translated from Japanese).

Response acceleration on the roof

The response acceleration on the roof compared to the ground is shown in Figure 7.8.

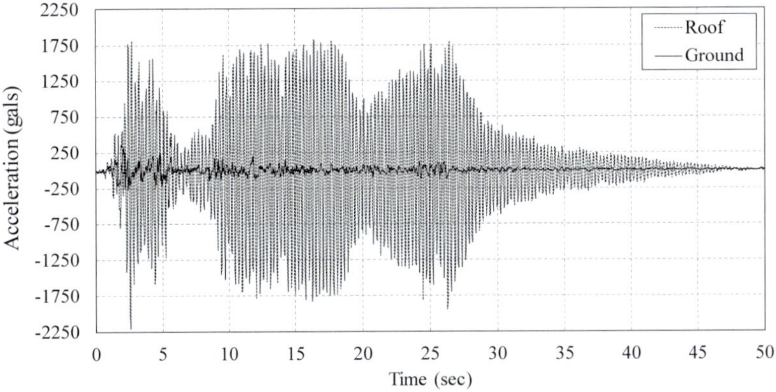

Figure 7.8 Response accelerations of the ground and roof of the building.

It is worth noting that the $A_{0.3}$ of the roof response acceleration is 1,739.9 gals (m-SIL=6.8) which is ten times larger than the original ground motion of 164.4 gals (m-SIL=5.2). Figure 7.9 shows how much the shaking of the ground (SIL=5 Strong) is amplified by shaking on the roof (SIL=7) of the building.

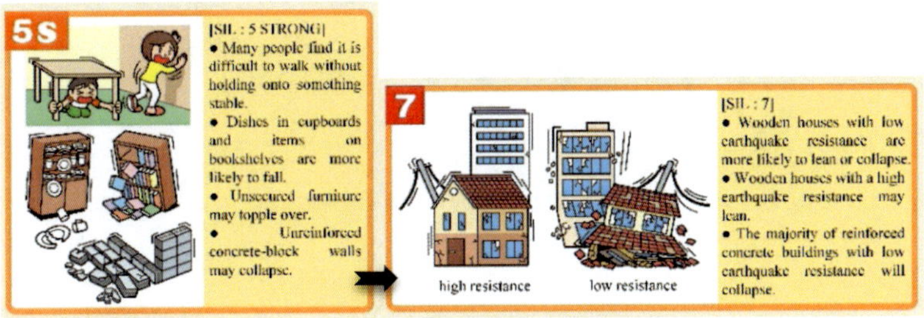

Figure 7.9 Illustrative representation of shaking quantity 7 of SIL.

The kinetic energy released on the roof

The time history and cumulative kinetic energy of the roof are shown in Figure 7.10. The kinetic energy is computed by the following equation.

$$E_k = \tfrac{1}{2} m_a \, v_a^2(t) \tag{7-1}$$

Where:

E_k is the kinetic energy of a particular floor in the building

m_a is the sum of the mass of the floor and half of the height of the columns

$v_a(t)$ is the velocity of the moving mass at a time t

This output will be used in the following sections to verify the effectiveness of various strengthening methods and show SIL's necessity as an evaluation tool.

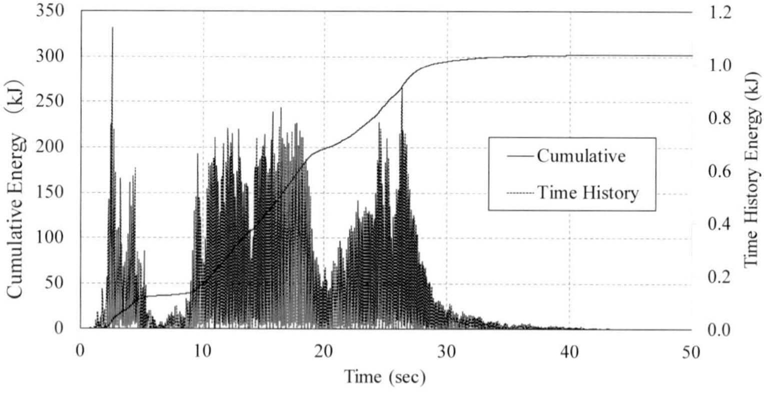

Figure 7.10 Cumulative kinetic energy on the roof of the building.

From Figure 7.10, the cumulative kinetic energy computed on the roof was

302.7 kJ (SIL=6.8). The mass of the roof was calculated from the sum of the beams and half of the height of the columns in the fifth story of the building to be 12,726.6 kg.

Condition assessment of an existing building

SIL values can be used to characterize the quantitative shaking levels in the case of a building in the existing condition. To do this, a building model of the existing building has to be modeled in a computer using the structural and properties data provided (Figure 7.11).

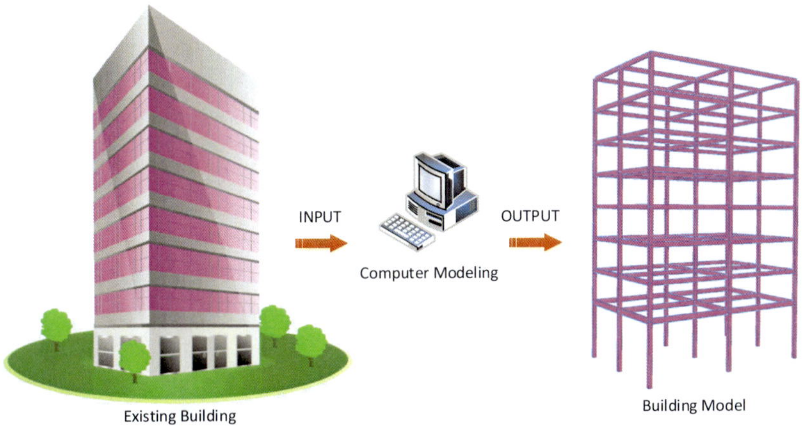

Figure 7.11 Structural modeling of a building using a computer.

Similar to the evaluation in the planning and design phases, to diagnose the condition of an existing building, using a computer to do numerical analysis, the shaking behaviors of the existing building can be estimated (Figure 7.12). From the estimation results, the building can be evaluated further on whether it needs to be strengthened. Therefore, this new seismic design philosophy will benefit practitioners who are building inspection consultants. Because the quantitative calculation of SIL will clearly describe the current building's shaking conditions during strong earthquakes.

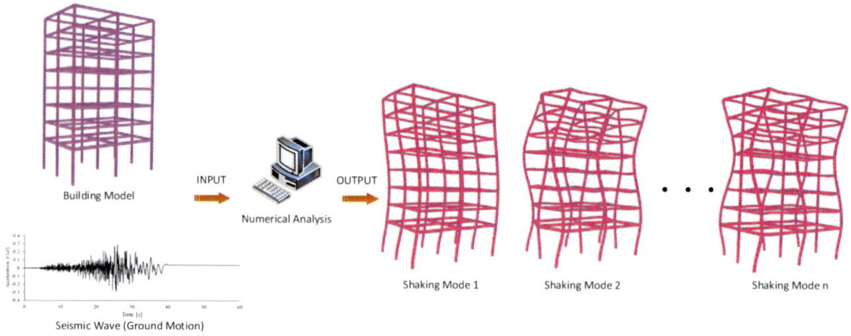

Figure 7.12 Prediction of shaking behavior of an existing building.

By seeing the SIL values on each floor in the existing building, as illustrated in Figure 7.13, we can use the SIL information for disaster mitigation preparedness before a strong earthquake occurs.

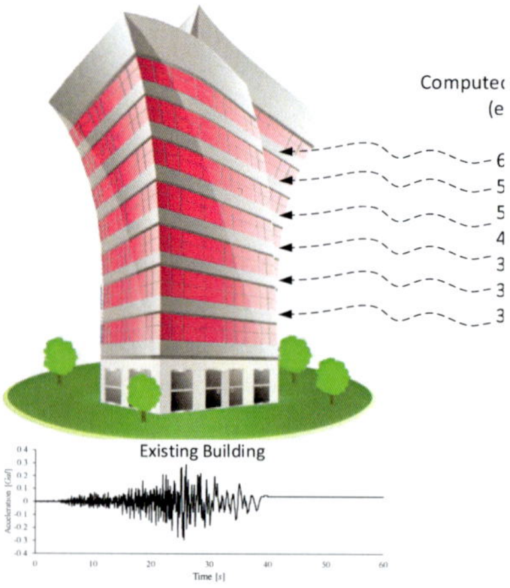

Figure 7.13 Prediction of building's SIL values on each floor for disaster mitigation.

Shaking potential of an existing building

In the case of an existing building, SIL evaluation can be a powerful tool to diagnose the shaking potential of the building at present. The existing building usually has design drawings. The building model can be easily constructed using the design drawings on a personal computer. Once the building model is ready, shaking using SIL can be evaluated.

Diagnosing the current potential shaking condition of the existing building during an earthquake can be achieved easily. From the evaluated SILs values on each floor of the building, we can estimate the shakings on each floor under a particular earthquake of concern.

Knowing the shaking characteristic of the floors in the building, a designer can easily determine what kind of retrofitting methods are necessary to reduce the shaking inside the building. All of the tasks can be done even on a single personal computer.

Real monitoring devices can be set to continuously record the waves to detect the responses of the floors in the existing building. The recorded acceleration waves can be used to produce the values of SIL on each floor inside the building.

The new technology for inspecting a building, so-called Non-Destructive Testing (NDT), can also be used to calibrate the existing building computer model by fitting the response variables measured at a particular location to the computed response variables such as displacement, velocity, or acceleration.

Strengthening a building against earthquakes

Among practices for strengthening buildings against earthquakes, bracings (active, passive) are the most adopted methods due to their economic considerations. Currently, the only evaluation procedure available in practice is to check if the strength criteria of the strengthened building meet the latest seismic design codes (Figure 7.14).

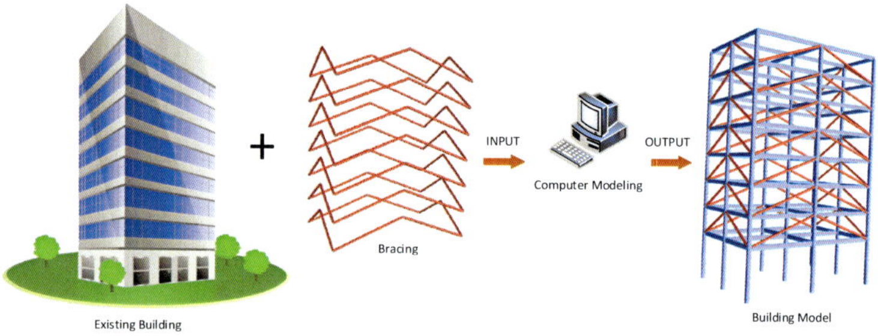

Figure 7.14 Structural modeling of a building strengthened by bracings.

Figure 7.15 shows how computer modeling is used to verify the effects of strengthening the building during earthquake occurrences.

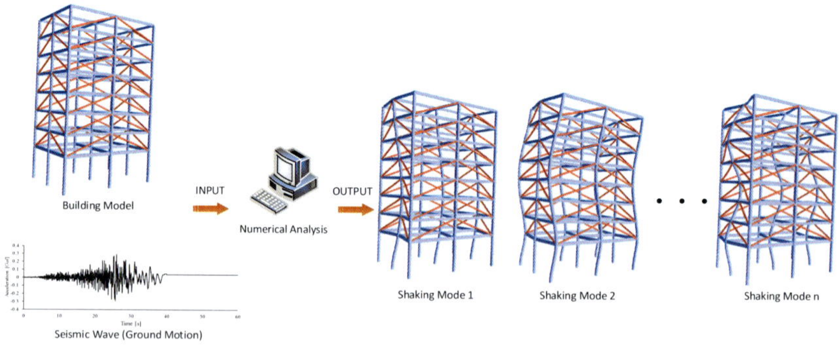

Figure 7.15 Verifying the performance of strengthening an existing building.

Even though the building is stronger after being strengthened, it cannot be guaranteed that it is free of shaking. This is because by making the building stiffer, its natural frequency will be shorter than before the strengthening. The shorter natural frequency of a building will make the building more prone to shaking due to strong earthquakes. The reason is that strong earthquakes often have a short dominant frequency in which the building will be at high risk of being shaken due to resonance phenomena.

Information about the SIL values of the building will make designers more confident in decision-making (Figure 7.16). Moreover, the SIL values can be used as a problem solver in helping a designer determines the type of strengthening method suitable for the building to reduce shaking.

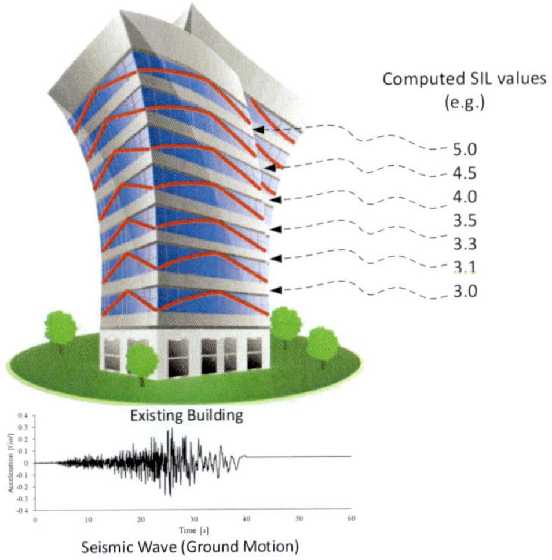

Figure 7.16 SIL as an evaluation tool for selecting a strengthening method for a building.

SILs application in strengthening methods

Whenever a significant earthquake stronger than previously ever recorded hits a certain area in a country, it is a common tradition that the current seismic design codes are revised. The newly revised codes usually recommend that the existing buildings designed using older codes need upgrades according to the newly revised codes.

However, the concept of "strength" is still the only criterion considered in the current design seismic codes. The new codes are revised by increasing the seismic loading to consider stronger earthquakes. Therefore, as long as the designers follow the revised seismic codes, the strengthening method adopted can be certified by the local governing institutions.

Even during the decision-making process, where many strengthening methods are considered candidates to be adopted, selecting the best and optimal strengthening method is only based on the economic aspect as the criterion.

The concept of "comfort," closely related to the shaking conditions of the building during an earthquake, is recommended to be augmented to the existing seismic design codes. Hence, using the SIL values as the code's additional design criterion will assist in finding the best solution among the strengthening methods.

The application of SILs values will be demonstrated in the following section to show the effectiveness of the quantitative-based selection procedure for the most appropriate strengthening method for the existing building. This procedure is not yet established in the current seismic design codes.

Various strengthening cases

As a demonstrative example, the design calculation based on the new revised seismic design codes requires the existing building to be strengthened by bracing types, which require at least 9.5% weight of the entire building structure. Table 7.3 shows four possible methods for strengthening by bracing the existing building.

Table 7.3 Strengthening methods use bracings of types I, II, III, and IV.

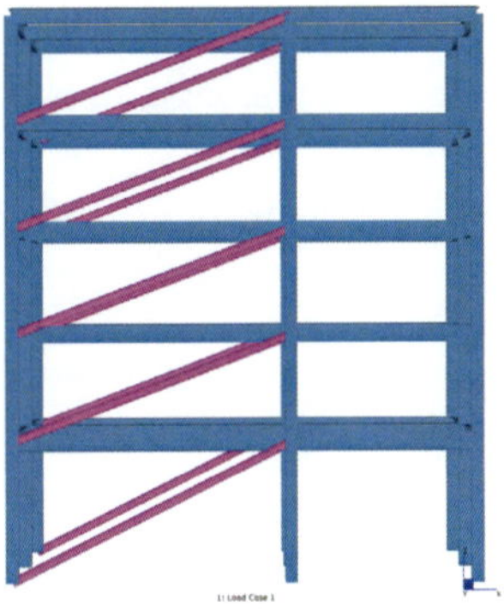

Type I (+9.5%Weight)

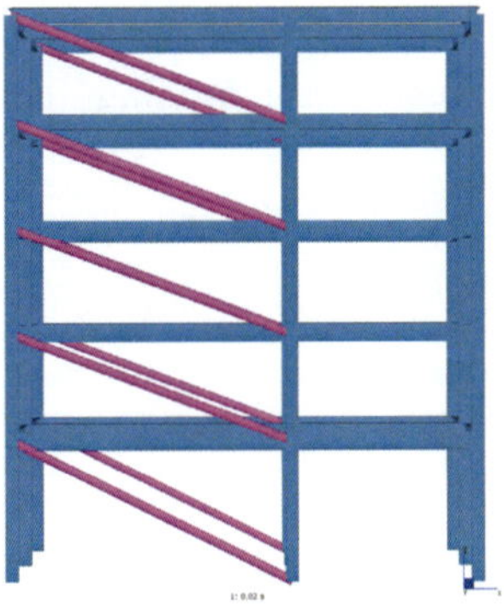

Type II (+9.5%Weight)

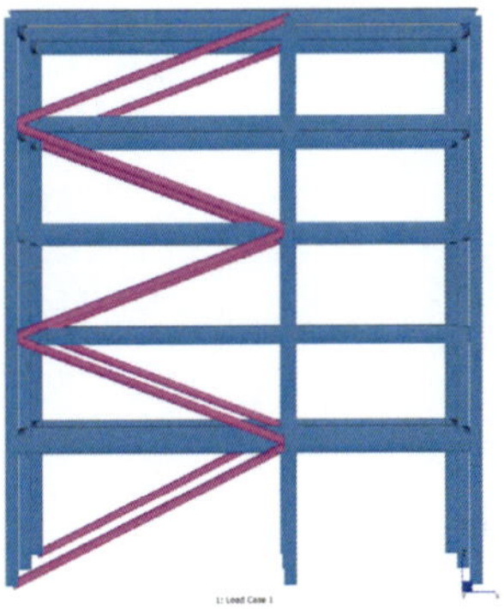

Type III (+9.5%Weight)

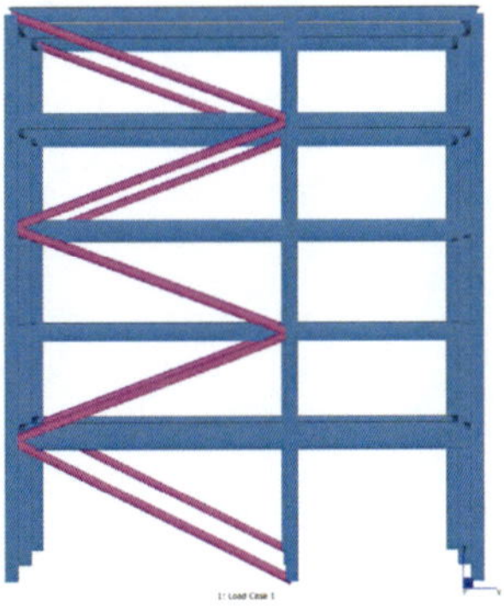

Type IV (+9.5%Weight)

In current design practice, these four strengthening bracing methods equally fulfill the requirements of the accredited strengthening methods.

What would the results be if the SIL evaluation concept is applied to these cases using the same El Centro earthquake as the ground motion? Different earthquake ground motions can be easily tested to see how these four bracing types perform. How much improvement was made in terms of SIL values in reducing the shaking of the existing building?

In real practice, where SIL is not used as an evaluation tool, no specific criteria can give a meaningful evaluation for decision-making purposes.

Table 7.4 shows the calculated m-SIL for the four strengthening methods of I-IV, as shown in Table 7.3.

Table 7.4 SILs of strengthening method types I–IV.

Floor	**Calculated m-SIL (+9.5%Weight Types)**				
	Existing	**I**	**II**	**III**	***IV***
Roof	6.8	6.6	6.6	6.6	***6.6***
5	6.7	6.5	6.5	6.5	***6.5***
4	6.5	6.3	6.3	6.3	***6.3***
3	6.2	5.9	5.9	5.9	***5.9***
2	5.8	5.5	5.5	5.5	***5.5***
Ground	5.2	5.2	5.2	5.2	***5.2***

From Table 7.4, the strengthening types I-IV show similar SIL values on each floor in the building. These four strengthening methods can be selected at liberty since they give the same m-SIL values on each floor. Although the strengthening methods reduce the shaking in the existing building, Table 7.4 shows that they only slightly reduce the shaking to a maximum of 0.3

m-SIL difference of values. These shaking reductions are not significant.

The earthquake early warning of JMA illustrates the shaking condition on the fifth floor. Figure 7.17 shows the shaking condition on the fifth floor of the existing building and strengthening types; – - IV, which can be categorized into level 7 (≥6.5) of SIL.

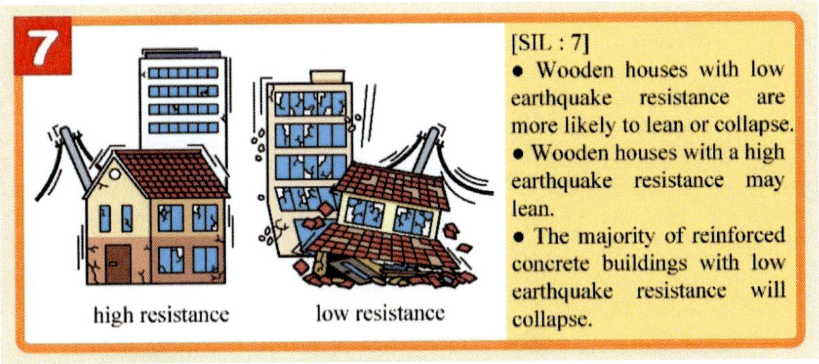

Figure 7.17 Illustrative representation of shaking quantity 7 of SIL.

Level 7 of the SIL is the maximum level to describe the shaking condition on the fifth floor and roof, supposing the El Centro 1941 earthquake struck. This example emphasizes that the use of SIL evaluation can reveal the 1940 El Centro earthquake occurrence physically. Even so, the earthquake only had a maximum ground acceleration of 342 gals and a Richter magnitude of 5.5, which is about one-third the gravitational acceleration but can cause significant damage to more than 80% of the buildings in the city. It is worth noting that the damage to the buildings at that time cannot be justified by only using maximum ground acceleration or energy.

Shaking performance evaluation of Bracing Type IV

The shaking performance evaluations of the building prototype strengthened by the representative Bracing Type IV with the 1940 El Centro as the input ground motion are shown in Table 7.5 and Figure 7.18. It can be seen in Table 7.5 that the $A_{0.3}$ accelerations and calculated m-SIL values, in general, are increasing in the height direction.

Table 7.5 Results of m-SIL of the building strengthened by Bracing Type IV.

Floor	m-SIL	T (sec) [FFT]	$A_{0.3}$(gals)	Natural Frequencies of Bracing Type IV
Roof	6.6	0.2165	1564.7	
5	6.5	0.2165	1394.5	Mode 1: T = 0.2165 sec
4	6.3	0.2165	1107.7	Mode 2: T = 0.1079 sec
3	5.9	0.2165	698.9	Mode 3: T = 0.1075 sec Mode 4: T = 0.0928 sec
2	5.5	0.2165	441.0	Etc.
Ground	5.2	0.6836	164.4	

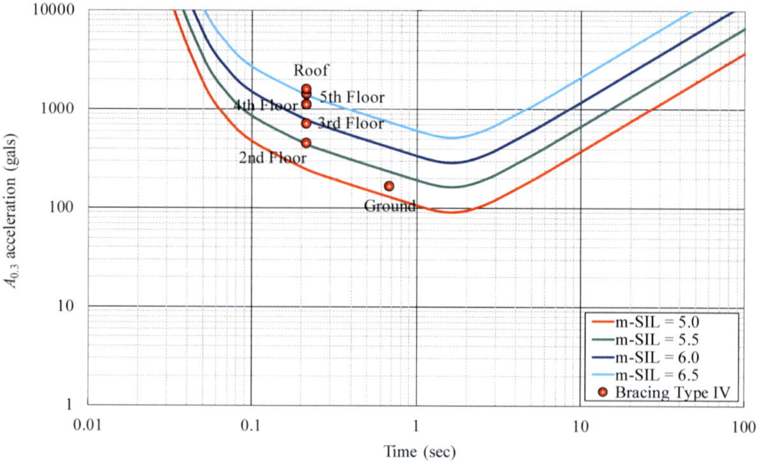

Figure 7.18 SIL of the building strengthened by Bracing Type IV.

As summarized in Table 7.5, generally, the m-SIL values on each floor in the building strengthened by Bracing Type IV are reduced slightly from the m-SIL values of the existing building before the strengthening. From the calculated m-SIL values, the maximum accelerations at each position can be computed following the procedure described in Chapter Five. By plotting the values of the dominant period and $A_{0.3}$ accelerations in the SIL graphs, we can see that the m-SIL at the ground surface is much less than on the roof (Figure 7.18). In applying the present methodology, Figure 7.18 should be used with Figure 7.7 as accompanying illustrative shakings of each floor in the building.

Effectiveness of Bracing Types I to IV

The response acceleration of the roof before and after strengthening with Bracing Type IV (the representative case) is shown in Figures 7.19 and 7.20.

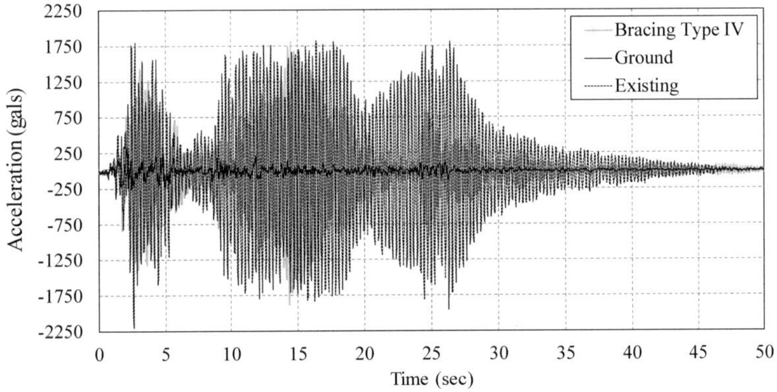

Figure 7.19 Response accelerations of the roofs of the buildings.

It is worth noting that the $A_{0.3}$ response accelerations of the roof before and after being strengthened are 1,739.9 gals (m-SIL=6.8) and 1,564.7 gals (m-SIL=6.6), respectively (Figure 7.19). The reduction of maximum acceleration is 175.2 gals (10.1% reduction).

From the energy dissipation point of view, the cumulative kinetic energies of the roof before and after being strengthened are 302.7 kJ (m-SIL=6.8) and 103.0 kJ (m-SIL=6.6), respectively (Figure 7.20).

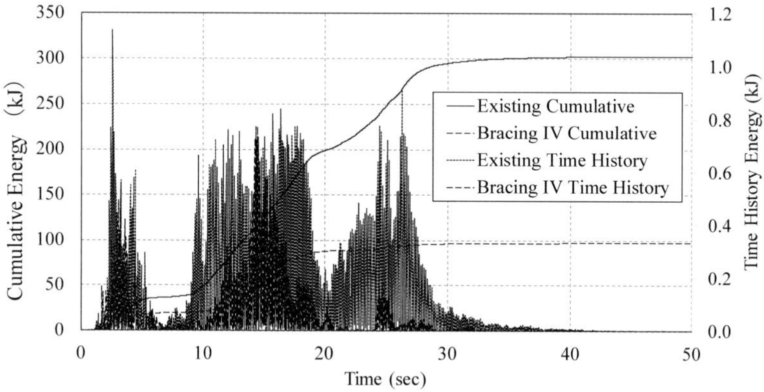

Figure 7.20 Kinetic energy of the roofs of the buildings.

The energy dissipated after the strengthening is 103.0 kJ (34.0%). It seems likely that the energy dissipated by the strengthening method is quite high because it can dissipate the kinetic energy close to 70% from the existing building without strengthening. However, the value of SILs in both cases does not show better performance in reducing the shaking of the roof of the building.

We can see here that the evaluation criteria using maximum acceleration or energy dissipation may lead to different conclusions for designers to use as a decision-making tool. However, the evaluation using SIL is more consistent and easier to understand. Before and after being strengthened, cases are within the 6.5 and 7.0 range of SIL and still possess a potential strong shaking in category level 7 of SIL.

Next, adding bracings of 16.3% (addition of 6.8%) was investigated by adding bracings to the right span of the frame structure of the building (Table 7.6).

Table 7.7 shows the results of the calculated SIL of the four strengthening methods of V-VIII, shown in Table 7.6.

From Table 7.7, the strengthening types V-VIII show similar SIL values on each floor in the building compared with the strengthening types I-IV. This tells us that the addition of 6.8% weight of bracings does not reduce the intensity of shakings on each floor in the building significantly, only by about 0.1 scales in SIL values. These shaking reductions are not worth considering because they will significantly increase the cost of materials. However, without knowing the present methodology, the strengthening method types V-VIII will gain favor among designers.

Table 7.6 Strengthening methods use of bracings of types V and VIII.

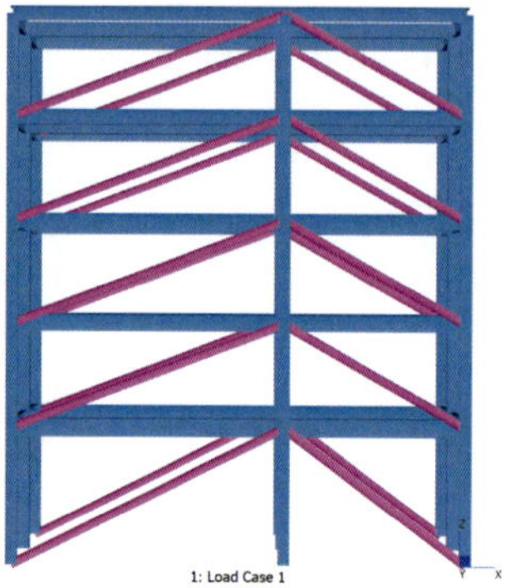

Type V (+16.3%Weight)

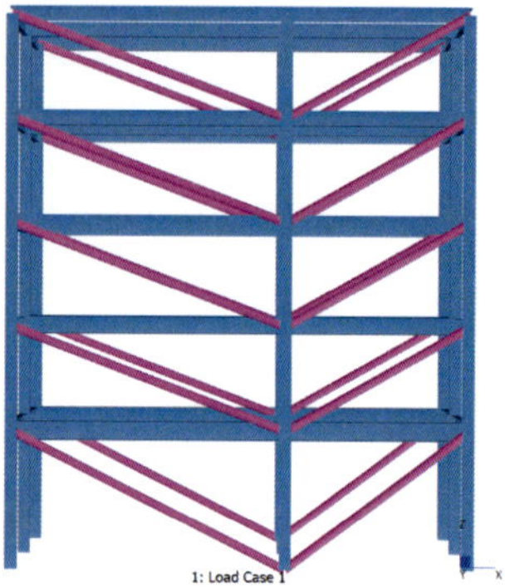

Type VI (+16.3%Weight)

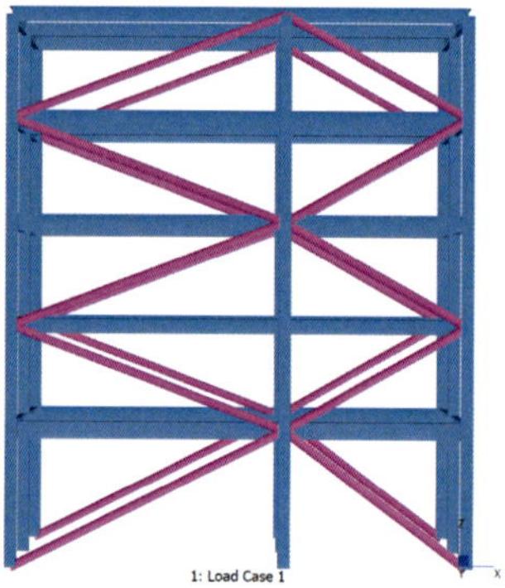

Type VII (+16.3%Weight)

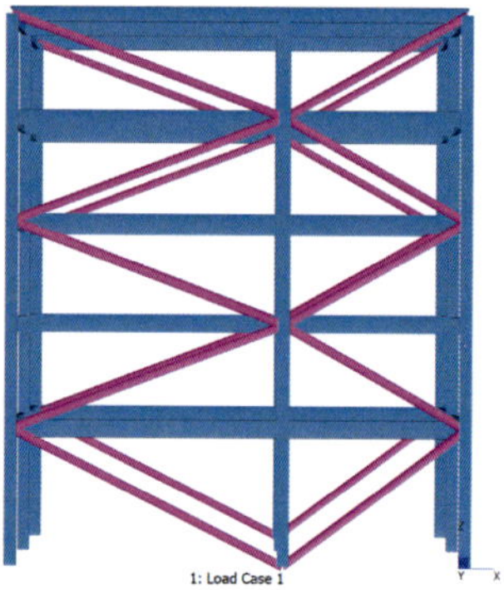

Type VIII (+16.3%Weight)

Table 7.7 SILs of strengthening method types V–VIII.

Floor	Calculated SIL (+16.3%Weight)				
	Existing	V	VI	VII	VIII
Roof	6.8	6.5	6.5	6.5	6.5
5	6.7	6.4	6.4	6.4	6.4
4	6.5	6.2	6.2	6.2	6.2
3	6.2	5.9	5.9	5.9	5.9
2	5.8	5.5	5.5	5.5	5.5
Ground	5.2	5.2	5.2	5.2	5.2

Figure 7.21 shows the shaking condition on the fifth floor with strengthening types V-VIII that can be categorized into level 6 Strong (6.0–6.4) of SIL.

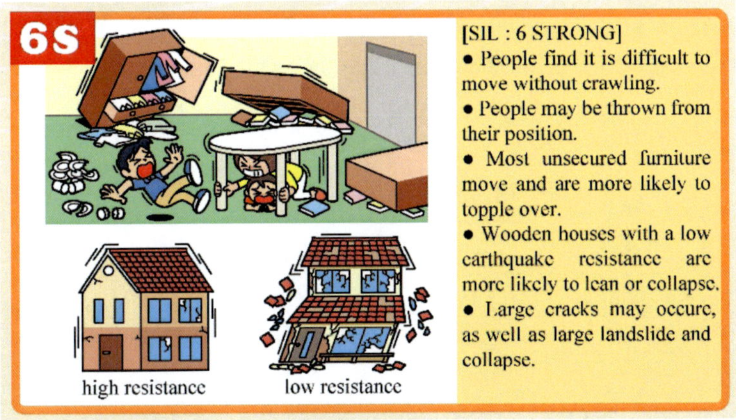

Figure 7.21 Illustrative representation of shaking quantity 6 Strong of SIL.

Finally, the addition of cross bracing types IX-X of 18.9% and 32.6% (addition of 9.4% and 23.1%) to the left span only and the full span of the frame of the building are shown in Table 7.8.

Table 7.8 Strengthening methods use of bracings of types IX and X.

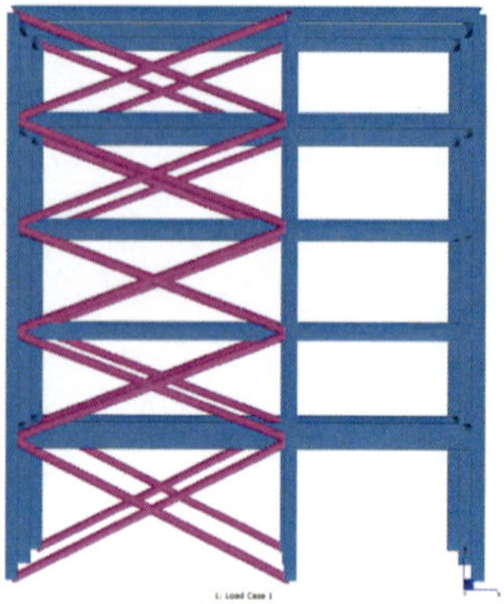

Type IX (18.9%)

Type X (32.6%)

Table 7.9 shows the results of calculated m-SIL for the two strengthening methods of IX and X, as shown in Table 7.8.

Table 7.9 SILs of strengthening method types IX and X.

Floor	**Calculated m-SIL**		
	Existing	IX (18.9%)	***X (32.6%)***
Roof	6.8	6.5	***6.4***
5	6.7	6.4	***6.3***
4	6.5	6.2	***6.1***
3	6.2	5.9	***5.8***
2	5.8	5.5	***5.5***
Ground	5.2	5.2	***5.2***

From Table 7.9, the strengthening types IX-X show fewer improvements to the m-SIL values on each floor in the building. Compared with the strengthening types I-VIII, adding either 9.4% (type IX) or 23.1% (type X) weight of bracings does not reduce the shaking on each floor in the building considerably but by about only 0.3-0.4 scale of m-SIL values reduction.

Figure 7.21 shows the shaking condition on the fifth floor and roof of the strengthening types IX-X can be categorized into level 6 Strong (6.0–6.4) and 7 (≥6.5) of SIL.

It can be noted that even though the bracings are increased, there is no improvement in reducing the shaking of the building significantly because the building is still shaking at a maximum of level 7 in the occurrence of an earthquake resembling the 1940 El Centro earthquake.

Shaking performance evaluation of Bracing Type X

The shaking performance evaluations of the building prototype strengthened by the representative Bracing Type X with the 1940 El Centro earthquake as the input ground motion are shown in Table 7.10 and Figure 7.22. It can be seen in Table 7.10 that the $A_{0.3}$ accelerations and calculated m-SIL values, in general, are increasing in the height direction.

Table 7.10 Results of m-SIL of the building strengthened by Bracing Type X.

Floor	m-SIL	T (sec) [FFT]	$A_{0.3}$(gals)	Natural Frequencies of Bracing Type IV
Roof	6.4	0.2158	1245.5	
5	6.3	0.2158	1110.0	Mode 1: T = 0.2141 sec
4	6.1	0.2158	881.7	Mode 2: T = 0.0928 sec Mode 3: T = 0.0690 sec
3	5.8	0.2158	624.2	Mode 4: T = 0.0690 sec
2	5.5	0.2158	441.9	Etc.
Ground	5.2	0.6836	164.4	

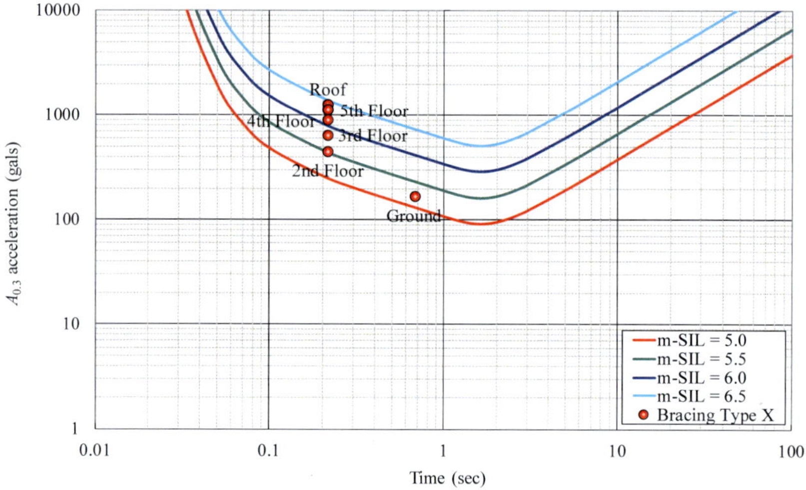

Figure 7.22 SIL of the building strengthened by Bracing Type X.

As summarized in Table 7.10, generally, the m-SIL values on each floor in the building strengthened by Bracing Type X are reduced slightly from the m-SIL values of the existing building before the strengthening. From the calculated m-SIL values, the maximum accelerations at each position can be computed following the procedure described in Chapter Five. By plotting the values of the dominant period and $A_{0.3}$ accelerations in the SIL graphs, we can see that the m-SIL at the ground surface is much less than on the roof (Figure 7.21). In applying the present methodology, Figure 7.21 should be used with Figure 7.7 as accompanying illustrative shakings of each floor in the building.

Effectiveness of Bracing Type X

The response acceleration of the roof before and after strengthening with Bracing type X (the most convincing case) is shown in Figures 7.23 and 7.24.

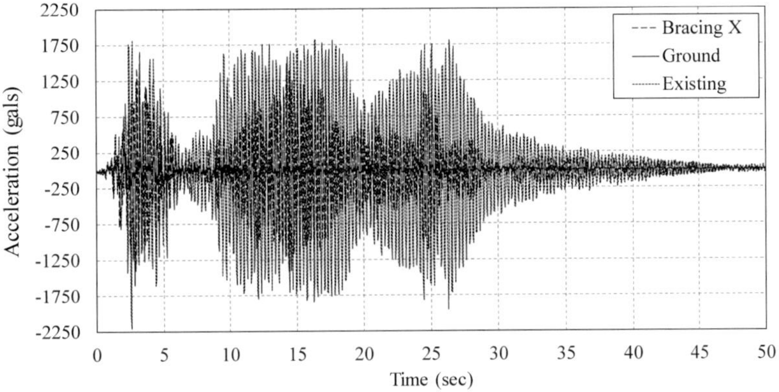

Figure 7.23 Response accelerations of the roofs of the buildings.

From Figure 7.23, the $A_{0.3}$ response accelerations of the roof before and after being strengthened are 1,739.9 gals (m-SIL=6.8) and 1,245.5 gals (m-SIL=6.4), respectively. The reduction of maximum acceleration is 494.4 gals (28.4%). It is noted that even though the building is strengthened by a full configuration of diagonal bracing, the reduction of maximum response acceleration is still not significant.

From the energy dissipation point of view, the cumulative kinetic energies of the roof before and after being strengthened are 302.7 kJ (m-SIL=6.8) and 100.0 kJ (m-SIL=6.4), respectively (Figure 7.24).

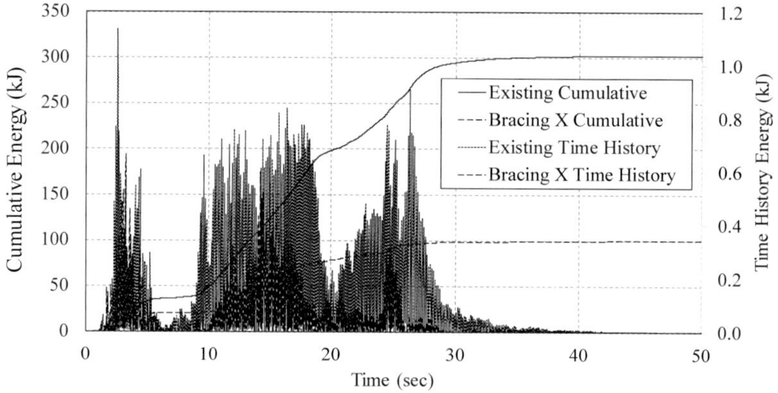

Figure 7.24 Kinetic energy of the roofs of the buildings.

The energy dissipated after the strengthening is 100.0 kJ (67.0%). It seems likely that the energy dissipated by the strengthening method is quite high because it can dissipate the kinetic energy of about 67% from the existing building without strengthening. However, the value of SILs in both cases shows slightly better performance in reducing the shaking of the roof of the building.

We can see here that the evaluation criteria using maximum acceleration or energy dissipation may lead to different conclusions for designers to use as a decision-making tool. However, the evaluation using SIL is more consistent and easier to understand, in which both before and after strengthening cases are reduced to only 0.4 in the range of SIL, which does not significantly reduce the shaking of the building.

The investigations on the strengthening using bracings have overviewed the SIL benefits in helping designers make an appropriate decision, especially when there is a necessary trade-off between the comfort of shaking reduction and the cost required to be spent. In real practice, many designers do not know what and how to decide based only on the strength criteria as recommended in the codes. As we can observe, in type X of the bracing method, the densest bracings placements are available. Still, they cannot significantly reduce the shaking quantity. Therefore, another strengthening method will be needed to find the optimal solution.

Reducing shaking in buildings from earthquakes

In the vibration control measures, SIL values can be used to measure the performance of earthquake-resistant instruments such as a base isolator, active/passive control devices, damper equipment, etc. SIL values can be beneficial for evaluating the outcomes of shaking control of a building. Many numerical investigations can be conducted to obtain the best and optimal performance with the lowest cost without conducting cost-consuming experimental work.

Seismic devices

SIL can be effectively used to investigate the performance of seismic devices such as Tune Mass Dampers (TMD), active suspension devices, or base isolators to be installed in a building (Figure 7.25).

Figure 7.25 SIL as a performance evaluation tool of seismic devices in a building.

In most common practices, designing seismic devices for a building is usually based on historical work experience data as the guidelines. However, it is more reliable and economical to evaluate SIL values using computer modeling of the building. The reason is that there is less possibility of designing two or more identical buildings, even though we have prototype designs of the buildings. Hence every building has to be modeled to a customized design. Moreover, major design changes to a part of the existing building will influence the devices' performance during a strong earthquake. In other words, any seismic device installed in a building must be designed to suit the characteristics of the building.

Conducting tests on these seismic devices is feasible only by using a smaller test model, which is scaled down from the original size of the building. Hence, having SIL values of the building installed by these seismic devices will help a designer to study and investigate the performances of seismic devices economically but verified mechanically.

Tuned Mass Damper device

A Tuned Mass Damper (TMD) is an active seismic device consisting of a mass, a spring, and a damper attached to a building, usually on the top of the highest floor or roof (Figure 7.26).

TMDs are also known as harmonic absorbers or seismic dampers. TMDs are used for anti-resonance of the resonant frequency of the building. The frequency of the TMD is tuned to be similar to the resonant frequency of the building so that when the building is excited due to a strong earthquake, the TMD will resonate out of the oscillation phase with the motion of the building. Seismic energy is dissipated by the damper inertia force acting on the building.

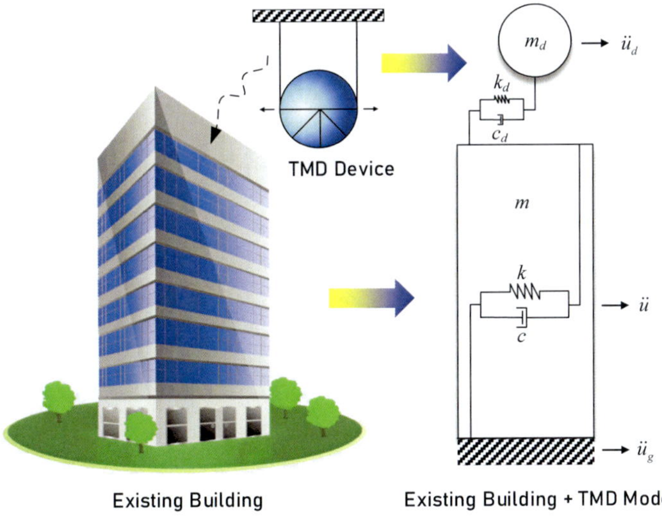

Figure 7.26 TMD vibration control system.

This section starts with implementing TMD for building vibration control by introducing TMD design and application examples. To assess the effectiveness of TMD, SIL calculation for each floor in the building is conducted. These SILs values of shakings are valuable information in decision-making. Numerous examples are provided to illustrate the effectiveness of vibration control of TMD devices which can be achieved under seismic excitation of a strong earthquake.

Generally, the TMD concept is illustrated using a two-mass system shown in Figure 7.26. Herewith, the subscript d refers to the TMD device. The structure of a building is idealized as a lump sum single mass system. The characteristic dynamic system of the building structure attached by a TMD is described by using the following notations.

$$\omega^2 = \frac{k}{m} \tag{7-2}$$

$$c = 2\,\zeta\,\omega\,m \tag{7-3}$$

$$\omega_d^2 = \frac{k_d}{m_d} \tag{7-4}$$

$$c_d = 2\,\zeta_d\,\omega_d\,m_d \tag{7-5}$$

where,

m is the entire mass of the building structure,

m_d is the mass of the TMD device,

k is the spring constant (stiffness) of the building structure,

k_d is the spring constant (stiffness) of the TMD device,

ω is the natural circular frequency of the building structure,

ω_d is the natural circular frequency of the TMD device,

ζ is the damping ratio of the building structure,

ζ_d is the damping ratio of the TMD device,

and defined $\overline{m}$ as the mass ratio,

$$\overline{m} = \frac{m_d}{m}$$

the governing equations of the building and TMD device attached are given by,

$$(1+\overline{m})\,\ddot{u} + 2\,\zeta\,\omega\,\dot{u} + \omega^2\,u = -\,\ddot{u}_g - \overline{m}\,\ddot{u}_d \tag{7-6}$$

and,

$$\ddot{u}_d + 2\,\zeta_d\,\omega_d\,\dot{u}_d + \omega_d^2\,u_d = -\ddot{u} \tag{7-7}$$

where,

u is the horizontal displacement of the building structure,

$\dot{u} = \dfrac{du}{dt}$ is the horizontal velocity of the building structure,

$\ddot{u} = \frac{d^2u}{dt^2}$ is the horizontal acceleration of the building structure,

u_d is the horizontal displacement of the TMD device,

$\dot{u}_d = \frac{du_d}{dt}$ is the horizontal velocity of the TMD device,

$\ddot{u}_d = \frac{d^2u_d}{dt^2}$ is the horizontal acceleration of the TMD device,

$\ddot{u}_g$ is the horizontal acceleration of the ground excitation,

t is the time.

In designing an optimal TMD to reduce the vibration of the building due to seismic excitation, the approximate optimal selection for the natural frequency of the TMD device is given as follows:

$$\omega_d \approx \omega \tag{7-8}$$

The spring constant of the TMD device is obtained from the following:

$$k_d = \overline{m}\, k \tag{7-9}$$

The relative maximum displacement of the TMD device $\hat{u}_d$ and building structure $\hat{u}$ can be defined by:

$$\hat{u}_d = \frac{1}{2\,\zeta_d}\hat{u} \tag{7-10}$$

In design practice, the equivalent damping ratio ε_e is given by:

$$\varepsilon_e = \frac{\bar{m}}{2}\sqrt{1+\left(\frac{2\zeta}{\bar{m}}+\frac{1}{2\zeta_d}\right)^2} \tag{7-11}$$

Assuming $\left(\frac{1}{2\zeta_d}\right) \gg \left(\frac{2\zeta}{\bar{m}}\right)$ and substituting ζ_d from Equation (7-10) into Equation (7-11) results in:

$$\bar{m} \approx 2\,\varepsilon_e \left(\frac{\hat{u}}{\hat{u}_d}\right) \tag{7-12}$$

Substitution of Equation (7-10) yields:

$$\bar{m} \approx 4\,\varepsilon_e\,\zeta_d\,. \tag{7-13}$$

Suppose we want to add a TMD device such that the equivalent damping ratio is 10% $(\varepsilon_e = 0.1)$ and the relative maximum displacement between TMD and building structure is 10 times $(\zeta_d = 0.05)$. From Equation (7-13), TMD's mass, spring constant, and damping ratio can be designed as follows.

$$\begin{aligned} \bar{m} &= 0.02\,(2\%) \\ k_d &= 0.02\,k \\ \zeta_d &= 0.05\,(5\%) \end{aligned}$$

Application of SIL in determining optimal TMD devices

Table 7.11 shows the parameter study of TMD devices by varying the mass, stiffness, and damping.

Table 7.11 Varying the properties of TMD devices from T01 to T10.

Case	Mass		Stiffness		Damping	
	m_d	%	k_d	%	c_d	ζ_d (%)
T01	1,466.63	2	1.1268	2	0.0352	5
T02	3,666.57	5	2.8171	5	0.0879	5
T03	7,333.13	10	5.6342	10	0.1758	5
T04	14,666.26	20	11.2683	20	0.3516	5
T05	21,999.39	30	16.9025	30	0.5274	5
T06	21,999.39	30	11.2683	20	0.5274	5
T07	21,999.39	30	28.1708	50	0.5274	5
T08	21,999.39	30	11.2683	20	1.5821	15
T09	21,999.39	30	28.1708	50	1.5821	15
T10	21,999.39	30	28.1708	50	2.6368	25

Table 7.12 shows the results of calculated SIL for the ten TMD devices of T01-T10.

Table 7.12 m-SILs of varying properties of TMD devices of T01–T10.

Floor	Calculated m-SIL										
	Existing	T01	T02	T03	T04	T05	T06	T07	T08	T09	***T10***
Roof	6.8	6.6	6.4	6.3	6.1	5.9	6.0	5.9	5.8	5.8	***5.8***
5	6.7	6.5	6.3	6.2	6.0	5.9	5.9	5.8	5.8	5.7	***5.7***
4	6.5	6.3	6.2	6.0	5.9	5.8	5.9	5.7	5.6	5.7	***5.6***
3	6.2	6.0	5.9	5.8	5.7	5.6	5.6	5.6	5.5	5.5	***5.5***
2	5.8	5.6	5.6	5.5	5.5	5.4	5.5	5.4	5.4	5.4	***5.4***
Ground	5.2	5.2	5.2	5.2	5.2	5.2	5.2	5.2	5.2	5.2	***5.2***

The shaking condition on the fifth floor of the existing building and using TMD types T–1 - T10 can be categorized into the levels of 6 Weak (T05–T10 are within 5.5–5.9, see Figure 7.27), 6 Strong (T02–T04 are within 6.0–6.4, see Figure 7.21) and 7 (T01 ≥ 6.5 , see Figure 7.17) of SIL.

Figure 7.27 Illustrative representation of shaking quantity 6 Weak SIL.

Shaking performance evaluation of TMD type 10

The shaking performance evaluations of the building prototype attached by a TMD device on the roof of type 10 with the 1940 El Centro as the input ground motion are shown in Table 7.13 and Figure 7.28. It can be seen in Table 7.13 that the $A_{0.3}$ accelerations and calculated m-SIL values, in general, are increasing in the height direction.

Table 7.13 Results of m-SIL of the building attached by TMD type 10.

Floor	m-SIL	T (sec) [FFT]	$A_{0.3}$(gals)	Natural frequencies of TMD type 10
Roof	5.8	0.2290 (0.6896)	600.9 (326.5)	
5	5.7	0.6896 (0.2290)	291.0 (535.6)	Mode 1: T = 0.6707 sec Mode 2: T = 0.2421 sec Mode 3: T = 0.2173 sec Mode 4: T = 0.1744 sec Etc.
4	5.6	0.2290 (0.6896)	477.3 (259.6)	
3	5.5	0.6896 (0.2290)	231.1 (425.4)	
2	5.4	0.6896 (0.2643)	206.0 (379.2)	
Ground	5.2	0.6836	164.4	

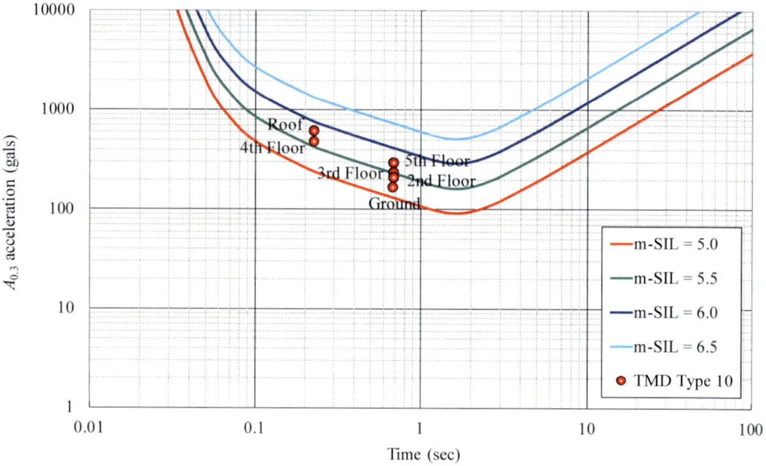

Figure 7.28 SIL values of the building attached by TMD type 10 on the roof.

As summarized in Table 7.13, in general, the m-SIL values on each floor of the building attached by TMD type X are reduced slightly from the m-SIL values of the existing building before the strengthening. In the case of attaching the TMD device on the roof as the seismic method, there is an interesting mechanism of the response of the building against the El Centro earthquake. As shown in Table 13, the dominant period and $\boldsymbol{A_{0.3}}$ acceleration values are given values inside the brackets that show another set of values (dominant period and $A_{0.3}$ accelerations) that are very close after conducting the FFT analyses of the response accelerations on each floor. This implies that whenever an earthquake has similar characteristics to the El Centro 1940, the assumed building prototype built on similar ground properties of soils will give two close behaviors of shakings to the building during the dominant period of 0.2290 and 0.6896 seconds. However, even though the dominant period of shaking was different, the building floors experienced the same quantity of m-SILs.

From the calculated m-SIL values, the maximum accelerations at each position can be computed following the procedure described in Chapter Five. By plotting the values of the dominant period and $A_{0.3}$ accelerations in the SIL graphs, we can see that the m-SIL at the ground surface is much less than on the roof (Figure 7.28). In applying the present methodology, Figure 7.28 should be used with Figure 7.7 as accompanying illustrative shakings of each floor in the building.

Effectiveness of TMD devices of Type T10

The response acceleration of the roof before and after using TMD of Type T10 installation (the most convincing case) is shown in Figures 7.29 and 7.30.

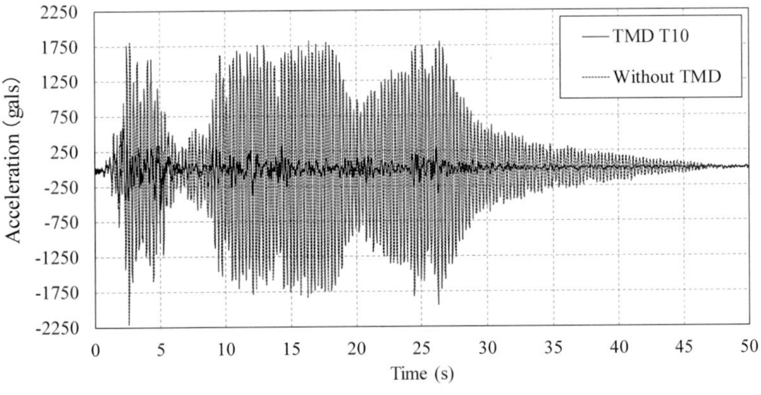

Figure 7.29 Response accelerations of the roof with T10 TMD compared with the roof without TMD.

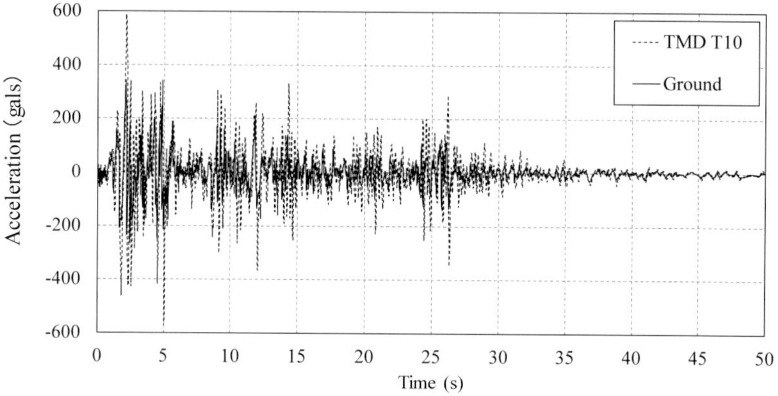

Figure 7.30 Response accelerations of the roof with TMD compared with ground acceleration.

From Figure 7.29, the $A_{0.3}$ response accelerations of the roof before and after the installation of TMD type T10 are 1,739.9 gals (m-SIL=6.8) and 600.9 gals (m-SIL=5.8), respectively. The reduction of $A_{0.3}$ response acceleration is 1,139.0 gals (65.5%). The reduction is very high. However, the value of SIL is only going down by one level below (i.e., from 6.8 to 5.8). It can be noted that the decreasing $A_{0.3}$ response acceleration at the roof does not decrease the value of m-SIL effectively.

Moreover, as shown in Table 7.11, to achieve maximum response acceleration reduction, the TMD with type T05–T10 should be made from a 30% mass (21,999.4 kg) of the total mass of the building structure. However, this massive weight of the TMD requires more space and a strong foundation on the roof of the building. Further detailed design of the foundation is necessary to ensure the safety of the entire structure of the building due to additional TMD weight.

From the energy dissipation point of view, the cumulative kinetic energies of the roof before and after installation of TMD type T10 are 302.7 kJ (SIL=6.8) and 13.5 kJ (SIL=5.8), respectively (Figure 7.31). The energy dissipated after the strengthening is 289.2 kJ (95.5%).

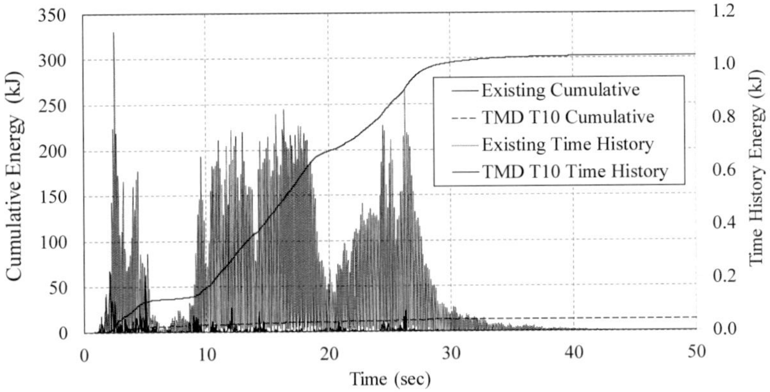

Figure 7.31 Kinetic energy of the roofs of the buildings with and without TMD.

It appears that the energy dissipated by using TMD of type T10 is very high because it can dissipate more than 95% kinetic energy from the existing building without TMD. However, the value of m-SILs in both cases shows slightly better performance in reducing the shaking of the roof of the building. The energy reduction is very high. However, the value of SIL is only going down by one level below (i.e., from 6.8 to 5.8). It is worth noting that decreasing energy dissipation at the roof does not decrease the value of m-SIL. The maximum reduction of m-SIL in these TMD types has a limit of 1.0 levels of shaking, which does not significantly reduce the shaking of the building. Therefore, if reducing the shaking is a prime factor in determining how to improve the existing building, another method can be needed to meet the shaking reduction purpose.

We can observe here, as done in most real aseismic design practices, the evaluation criteria by using maximum response acceleration or energy dissipation concept is not appropriate to be used as a tool to evaluate the reduction of shaking quantity of the building. However, the evaluation by using SIL is more consistent and easier to understand. Evaluation of the effectiveness of TMD by using SIL is recommended for designers to consider as an invaluable decision-making tool.

Investigations on building improvement using TMD devices have overviewed the important perspective of SIL's benefit in helping designers make an appropriate decision, especially when there is a necessary trade-off consideration between the comfort of shaking reduction and the cost required to be spent. In real practice, many designers do not know what and how to decide based only on the strength criteria as recommended in the codes.

Base Isolation device

A Base Isolation or Isolator (BI) device is a passive seismic device consisting of springs and dampers attached to the foundation of a building, usually under the basement floor of the building. (Figure 7.32).

BI is also known as a seismic absorber or seismic damper. The BI is used as an anti-resonance of the resonant frequency of the building. The frequency of the BI is tuned to be much smaller than the resonant frequency of the building so that when the building is less excited due to a strong earthquake, the BI will resonate out of the oscillation phase with the motion of the building. Seismic energy is dissipated by the damper inertia force acting on the building.

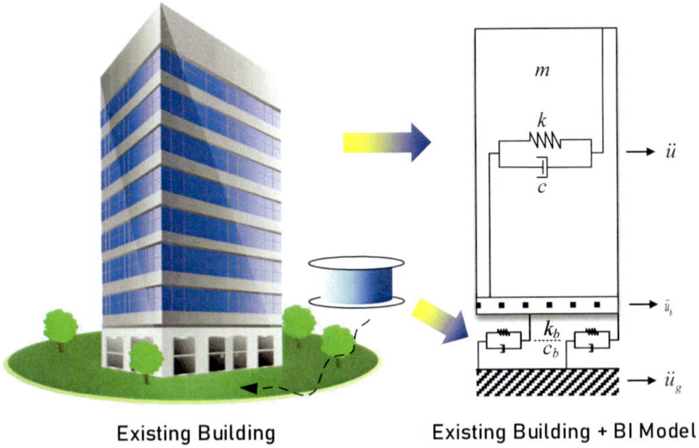

Figure 7.32 BI seismic isolation control system.

This section starts with implementing BI for vibration control in buildings by introducing an example of BI design and application. To assess the effectiveness of BI, SIL calculation on each floor in the building is conducted. These SILs values of shakings are valuable information in decision-making. Numerous examples are provided to illustrate the effectiveness of vibration control of BI devices which can be achieved under seismic excitation of a strong earthquake.

Generally, the concept of BI is illustrated by using a two-mass system, shown in Figure 7.32. Herewith, the subscript d refers to the BI device. The structure of a building is idealized as a lump sum single mass system. The characteristic dynamic system of the building structure attached by a BI is described using the following notations.

$$\omega^2 = \frac{k}{m} \tag{7-14}$$

$$c = 2\,\zeta\,\omega\,m \tag{7-15}$$

$$\omega_b^2 = \frac{k_b}{m_b} \tag{7-16}$$

$$c_b = 2\,\zeta_b\,\omega_b\,m_b \tag{7-17}$$

Where,

m is the entire mass of the building structure,

m_b is the mass of the slab foundation,

k is the spring constant (stiffness) of the building structure,

k_b is the spring constant (stiffness) of the BI device,

ω is the natural circular frequency of the building structure,

ω_b is the natural circular frequency of the slab foundation,

ζ is the damping ratio of the building structure,

ζ_b is the damping ratio of the BI device,

the governing equations of the building and base slab foundation are given by,

$$\ddot{u} + 2\,\zeta\,\omega\,\dot{u} + \omega^2\,u = -\ddot{u} \tag{7-18}$$

and,

$$(m+m_b)\,\ddot{u}_b + 2\,\zeta_b\, m_b\, \omega_b\, \dot{u}_b + m_b\, \omega_b^2\, u_b = -m_b\, \ddot{u}_g - m\,\ddot{u} \tag{7-19}$$

where,

u is the horizontal displacement of the building structure,

$\dot{u} = \dfrac{du}{dt}$ is the horizontal velocity of the building structure,

$\ddot{u} = \dfrac{d^2u}{dt^2}$ is the horizontal acceleration of the building structure,

u_b is the horizontal displacement of the slab foundation,

$\dot{u}_b = \dfrac{du_b}{dt}$ is the horizontal velocity of the slab foundation,

$\ddot{u}_b = \dfrac{d^2u_b}{dt^2}$ is the horizontal acceleration of the slab foundation,

$\ddot{u}_g$ is the horizontal acceleration of the ground excitation,

t is the time.

In designing an optimal BI to reduce shaking of the building due to ground excitation, the approximate optimal selection for the natural frequency of the BI device is given as,

$$\omega_b \;\square\square\; \omega \tag{7-20}$$

The natural frequency of the BI device is determined by the mass of the slab foundation on which the spring and damper are attached.

In practice, there are two basic elements in any practical seismic isolation system such as:

1. A damper is used to isolate the slab foundation from the ground by dissipating kinetic energy released by the ground motion during an earthquake.

2. A spring is used to isolate the slab foundation from the ground by lengthening the period of shaking using a flexible material such as steel or lead. However, the spring should not be too flexible to avoid excitation due to wind or small earthquakes.

A combination of elastomeric (lead-rubber) bearings with springs appears to be one of the most practical base isolators used in practice (Figure 7.33).

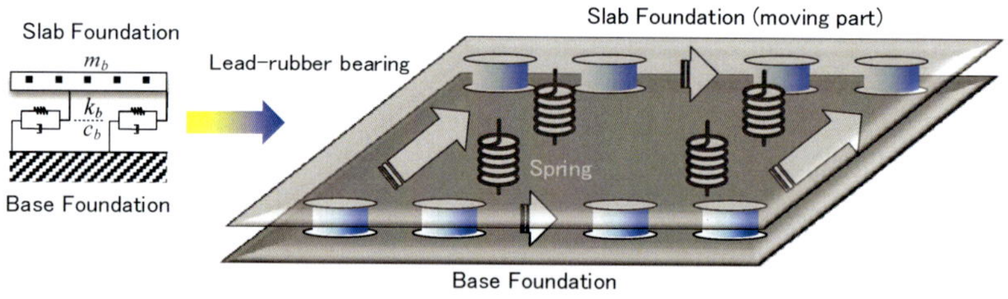

Figure 7.33 Base Isolator seismic isolation control system.

It should be noted that the base isolator devices were placed between the ground and the slab foundation. Therefore, the horizontal movements of the slab and ground are conceptually separated by the springs and dampers in between.

Application of SIL in determining optimal base isolator

Figure 7.34 shows the standard mechanical properties of a base isolator foundation. The base floor foundation is considered the mass of the base isolator. The base floor foundation is a reinforced concrete slab with a compressive design strength of 25 MPa. The slab is assumed to have a thickness of 30 cm. The mass of the moving slab is about 129,600 kg. Table 7.14 shows the parameter study of the base isolator by varying the stiffness and damper. There are nine devices installed under the concrete slab. The first natural frequency of the building is used to be compared with the base isolator's natural frequency.

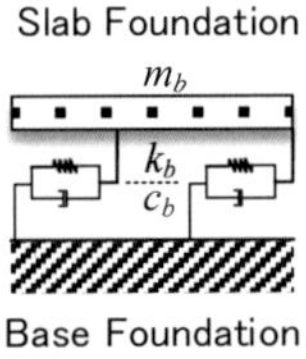

Figure 7.34 Mechanical properties of a base isolator.

Table 7.14 Varying the mechanical properties of Base Isolators B01 to B06.

Case	Stiffness of 9 devices		Damping of 9 devices		Frequency ratio
	k_{BI}	$\%k_{BUILDING}$	c_b	$\zeta_b(\%)$	$\omega_b / \omega\,(\%)$
B01	2.8170	2.9	15.5331	25	2.8
B02	1.1268	1.5	15.5331	25	2.8
B03	2.8170	2.9	6.9037	100	4.2
B04	1.1268	1.5	0.3452	5	1.3
B05	0.5634	0.58	0.0690	1	0.6
B06	***0.1125***	***0.12***	***0.0138***	***0.2***	***0.3***

Table 7.15 shows the calculated m-SIL of the six Base Isolators of Type B01–B06.

Table 7.15 m-SILs of varying properties of springs and dampers of B01–B06.

Floor	Calculated m-SIL						
	Existing	B01	B02	B03	B04	B05	***B06***
Roof	6.8	4.3	4.7	5.3	4.2	3.5	***2.3***
5	6.7	4.3	4.7	5.3	4.2	3.5	***2.3***
4	6.5	4.2	4.7	5.2	4.2	3.5	***2.2***
3	6.2	4.1	4.6	5.2	4.2	3.5	***2.2***
2	5.8	4.1	4.6	5.1	4.2	3.5	***2.2***
Ground	5.2	5.2	5.2	5.2	5.2	5.2	***5.2***

Table 7.15 shows that installing a base isolator can reduce the m-SIL values on the roof by more than four levels (6.8 to 2.3) from the m-SIL of the existing building roof.

Despite the material and maintenance costs of the base isolator, we can realize that, in general, by adjusting the mechanical properties of the springs and dampers, we can obtain the most reduction of shaking by using SIL as the evaluation criteria.

The shaking condition on the fifth floor of the existing building and using Base Isolator types B01–B06 varies from the levels of 2 (B06 is within 1.5–2.4, see Figure 7.35), 4 (B01, B04, and B05 are within 3.5–4.4, see Figure 7.36), 5 Weak (B02 is within 4.5–4.9) and 5 Strong (B03 is within 5.0–5.4, see Figure 7.37) of SIL.

Figure 7.35 Illustrative representation of shaking quantity 2 of SIL.

Figure 7.36 Illustrative representation of shaking quantity 4 of SIL.

Figure 7.37 Illustrative representation of shaking quantity 5 Weak and Strong of SIL.

Shaking performance evaluation of BI Type B06

The shaking performance evaluations of the building prototype with the installation of the most effective Base Isolators of Type B06 with the 1940 El Centro as the input ground motion are shown in Table 7.16 and Figure 7.38. It can be seen in Table 7.16 that the $A_{0.3}$ accelerations and calculated m-SIL values, in general, increase in the height direction.

Table 7.16 Results of m-SIL of the building with Base Isolator Type B06.

Floor	m-SIL	T (sec) [FFT]	$A_{0.3}$ (gals)	Natural Frequencies of BI Type B06
Roof	2.3	16.04	27.2	
5	2.3	16.04	27.2	
4	2.2	16.04	24.2	ω_{B06} =0.3917 rad/sec
3	2.2	16.04	24.2	T = 16.04 sec
2	2.2	16.04	24.2	
Ground	5.2	0.6836	164.4	

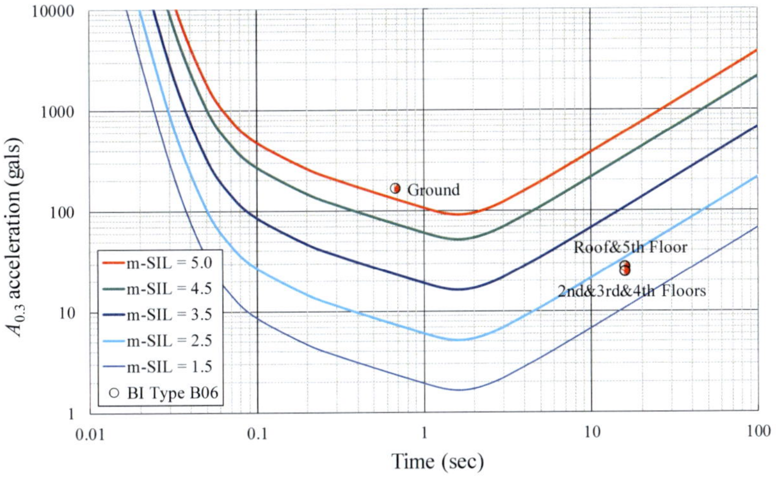

Figure 7.38 SIL of the building installed by BI Type B06 in the basement.

As summarized in Table 7.16, in general, the m-SIL values on each floor of the building installed with BI type B06 are reduced significantly from the m-SIL values of the existing building before the installation of BI devices. To obtain the most effective shaking reduction from a base isolator, the springs and dampers' natural vibration frequency must be very low (high period vibration) to hold the relationship ω_b □□ ω .

From the calculated m-SIL values, the $A_{0.3}$ accelerations on each position can be computed by following the procedure described in Chapter Five. By plotting the values of the dominant period and $A_{0.3}$ accelerations in the SIL graphs, we can see that the shaking intensity on every floor (m-SILl=2.2–2.3) in the building is small compared with the ground surface (Figure 7.38). In applying the present methodology, Figure 7.38 should be used with Figure 7.7 as accompanying illustrative shakings of each floor in the building.

Effectiveness of Base Isolators of Type B06

The response acceleration of the roof before and after installing Base Isolator type B06 (the most convincing case) is shown in Figures 7.39 and 7.40.

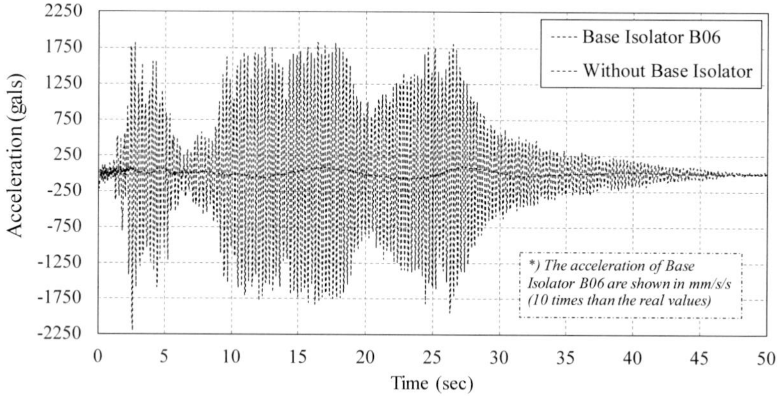

Figure 7.39 Response accelerations of the roof with Base Isolator B06 compared with the roof without a base isolator.

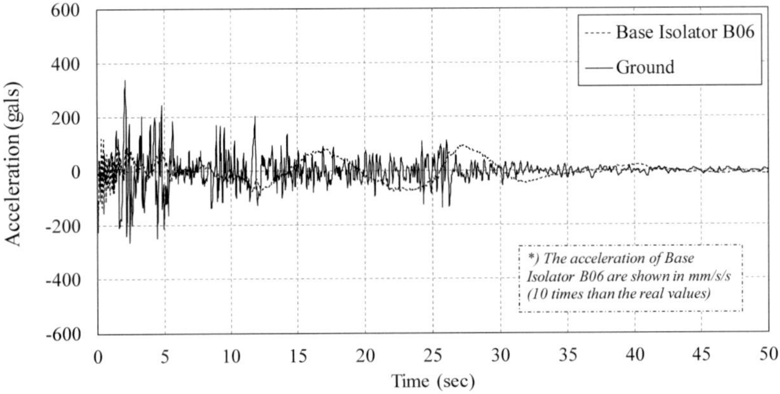

Figure 7.40 Response accelerations of the roof with Base Isolator B06 compared with ground acceleration.

From Figure 7.39, the maximum response accelerations of the roof before and after the installation of Base Isolator type B06 are 1,739.9 gals (m-SIL=6.8) and 27.2 gals (m-SIL=2.3), respectively. The reduction of maximum response acceleration is 1,712.7 gals (98.4%). Figure 7.17 shows the illustrative shaking of the existing building (SIL=7) before installing the base isolator. Figure 7.35 (SIL=2) illustrates the significance of the shaking reduction using the base isolator B06. The reduction factor is considerably higher than bracing strengthening methods and TMD devices. The value of SIL is reduced by about four levels (i.e., from 6.8 to 2.3). It is worth noting that the base isolator can decrease the maximum response acceleration of the roof and the value of SIL simultaneously.

From the energy point of view, the cumulative kinetic energies of the roof before and after installation of BI type B06 are 302.7 kJ (m-SIL=6.8) and 9.1 Joule (m-SIL=2.3), respectively (Figure 7.41). The energy dissipated after the installation of Base Isolator Type B06 is 293.6 kJ (97.0%).

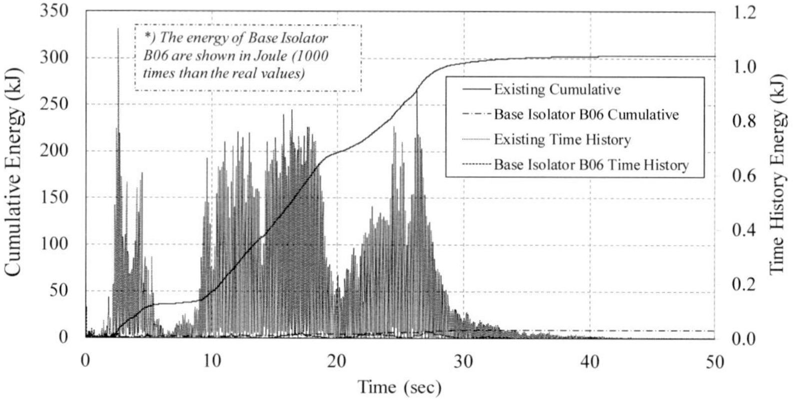

Figure 7.41 Kinetic energy of the roofs of the buildings.

From Figure 7.41, the energy dissipated by installing a base isolator, especially type B06, is very high because it can dissipate almost 100% of the kinetic energy from the existing building without the installation. This implies that the energy from the ground motion due to an earthquake is not transmitted into the building. The dampers of the base isolator play an important role in dissipating the energy.

TIe investigations on the aseismic effort using a base isolator have overviewed the important perspective of evaluation using SIL in helping designers make an appropriate decision, especially when there is a necessary trade-off consideration between the comfort of shaking reduction and the cost required to be spent. In real practice, many designers do not know how to decide based only on the strength criteria recommended in the seismic design codes. Most base isolator manufacturers usually plan and design a new building based on their work experiences in the past. There are no distinct criteria to evaluate the effectiveness of the installed base isolator. Therefore, the evaluation criteria of shaking, based on SIL with its illustrative quantitative descriptions of shaking, is highly recommended for real design practices.

Real-time monitoring

In a building, the information of SIL values in real-time measurement is used to alert occupants for disaster preparedness during a strong earthquake (Figure 7.42). Immediate results of the shaking level SIL can be used as real-time feedback to the local authority to collect disaster information on damage or injuries when a strong earthquake strikes a large administrative area. Collecting the disaster information of SILs makes it easy to identify the problem-solution process. Therefore, the local authority can take the necessary corrective disaster countermeasure actions by using real-time monitoring information of SILs. Real-time monitoring is also useful for avoiding existing and new disaster risks during strong earthquakes.

A risk assessment using SIL information can play a critical role in preparedness before a strong earthquake strikes or provide an initial and rapid estimation of human injuries and building damage.

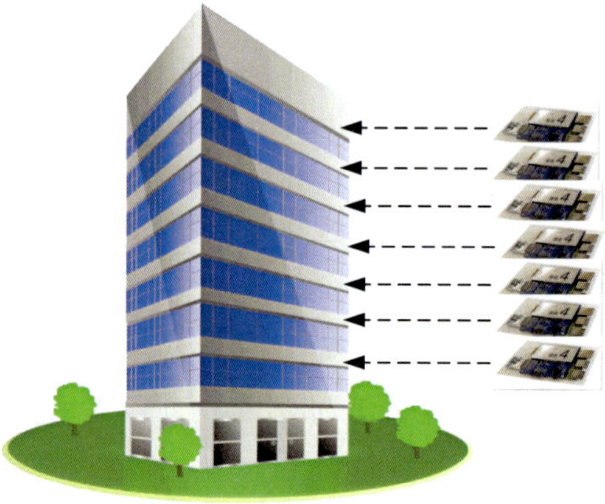

Figure 7.42 SIL for real-time monitoring device in a building.

Performance and cost evaluation

A cost-benefit analysis is necessary to facilitate designers' decision-making on implementing SIL as an evaluation tool for strengthening an existing building and installing aseismic devices. Because SIL is intended for existing seismic-resistant buildings already designed by the seismic codes, reducing shaking in terms of SIL can be considered an optional choice that can sometimes be costly. The cost can be considered a luxury purchase of the "comfort" of living in a building with a lower shaking level of SIL. This section will discuss the relationships between the cost needed for "buying" the low shaking evaluated by SIL. Finally, an indicative chart for the cost estimation of SIL will be provided for designers to use in their decision-making practices.

Cost estimation

Estimating seismic technology costs varies with materials, construction, maintenance, and others. Construction and maintenance costs are inherent to the native locality where the buildings are built. Only the material cost (in Japan, August 2022) will be considered in this section.

Table 7.17 Estimation of unit cost material.

Method	Type	Unit Cost/Ton	Note
Bracing	I~X	140	In 1,000 JPY, 2022
TMD	T01~T10	210	Estimated for a new building
Base Isolator	B01~B06	480	(facilities and maintenance)

The unit cost of TMD installation is approximately 1.5 times that of bracing methods by considering the cost of facilities and maintenance on the top floor of the building. The cost of elastomeric bearings for a base isolator is not easy to obtain from manufacturers because most of the costs are based on their working experiences in the past. Besides, every building has to be customized to meet the performance desired. Hence the cost also varies uncertainly. Herewith, the estimation of unit cost material is assumed by considering the advanced technology in a base isolator. Whenever there are updates in the unit cost among the aseismic methods, Table 7.17 can be modified.

Trade-off

A trade-off is a situational decision involving losing some of one quantity in return for gains in another. In this section, a trade-off is where either the cost or SIL value increases, and either the SIL value or cost must decrease.

The trade-off concept fits the situation where reducing the SIL value requires increasing the cost.

Indicative chart for cost estimation of SIL

A Pareto frontier solution is widely used in engineering to develop a cost estimation of SIL in a trade-off situation. The Pareto frontier allows the designer to restrict attention to an efficient choice by making a trade-off between the cost and SIL value.

Table 7.18 summarizes all the seismic technologies used as examples in this chapter.

Table 7.18 SIL values and total material cost estimation.

Type	SIL on each floor with its cost of building material					
	2	3	4	5	Roof	Cost in 1,000 JPY
I	5.5	5.9	6.3	6.5	6.6	972
II	5.5	5.9	6.3	6.5	6.6	972
III	5.5	5.9	6.3	6.5	6.6	972
IV	5.5	5.9	6.3	6.5	6.6	972
V	5.5	5.9	6.2	6.4	6.5	1,675
VI	5.5	5.9	6.2	6.4	6.5	1,675
VII	5.5	5.9	6.2	6.4	6.5	1,675
VIII	5.5	5.9	6.2	6.4	6.5	1,675
IX	5.5	5.9	6.2	6.4	6.5	1,940
X	5.5	5.8	6.1	6.3	6.4	3,347
T01	5.6	6.0	6.3	6.5	6.6	308
T02	5.6	5.9	6.2	6.3	6.4	770
T03	5.5	5.8	6.0	6.2	6.3	1,540
T04	5.5	5.7	5.9	6.0	6.1	3,080
T05	5.4	5.6	5.8	5.9	5.9	4,620
T06	5.5	5.6	5.9	5.9	6.0	4,620
T07	5.4	5.6	5.7	5.8	5.9	4,620
T08	5.4	5.5	5.6	5.8	5.8	4,620
T09	5.4	5.5	5.7	5.7	5.8	4,620
T10	5.4	5.5	5.6	5.7	5.8	4,620
B01	4.1	4.1	4.2	4.3	4.3	10,560
B02	4.6	4.6	4.7	4.7	4.7	10,560
B03	5.1	5.2	5.2	5.3	5.3	10,560
B04	4.2	4.2	4.2	4.2	4.2	10,560
B05	3.5	3.5	3.5	3.5	3.5	10,560
B06	2.2	2.2	2.2	2.3	2.3	10,560

Figure 7.43 shows the results of correlations between SIL values and various aseismic methods on the second floor of the building.

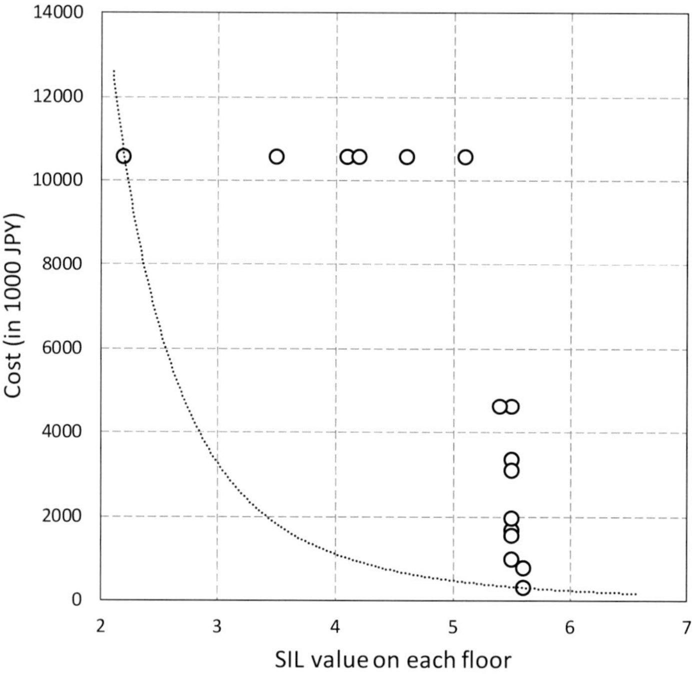

Figure 7.43 SIL and Cost of various aseismic methods on the second floor.

The trendline, which follows the power law of the optimal solution, is given as follows:

$$\text{Cost (in 1000 JPY)} = 208514 \times (\text{SIL})^{-3.783} \tag{7-21}$$

Figure 7.44 shows the results of correlations between SIL values and various aseismic methods on the third floor of the building.

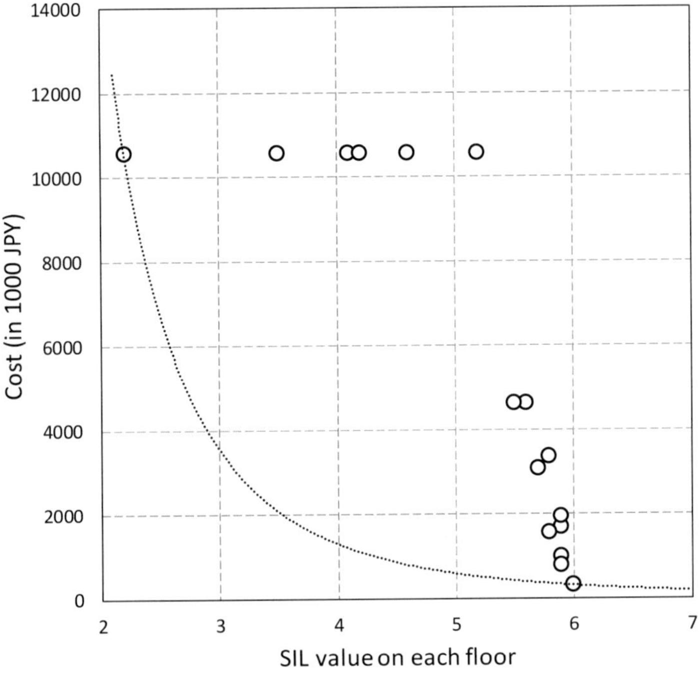

Figure 7.44 SIL and Cost of various aseismic methods on the third floor.

The trendline, which follows the power law of the optimal solution, is given as follows:

$$\text{Cost (in 1000 JPY)} = 169845 \times (\text{SIL})^{-3.523} \tag{7-22}$$

Figure 7.45 shows the results of correlations between SIL values and various aseismic methods on the fourth floor of the building.

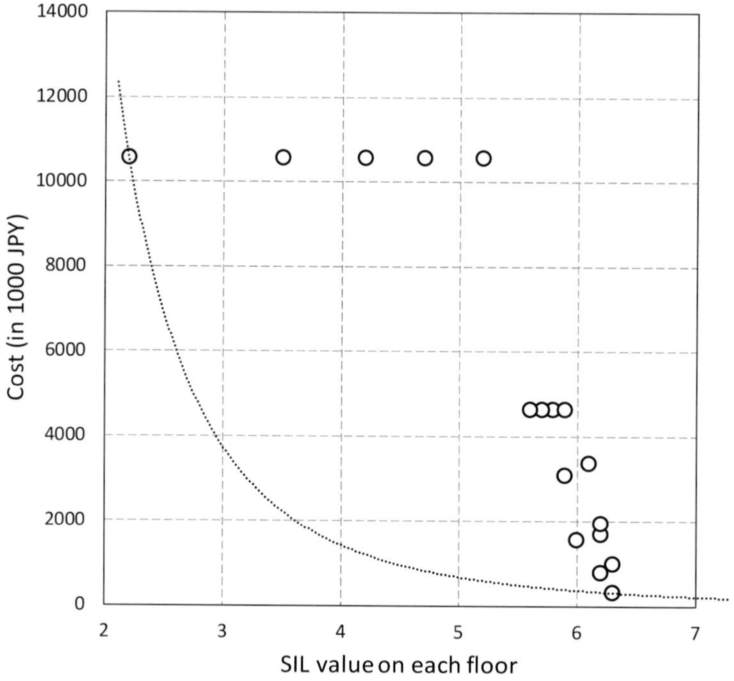

Figure 7.45 SIL and Cost of various aseismic methods on the fourth floor.

The trendline, which follows the power law of the optimal solution, is given as follows:

$$\text{Cost (in 1000 JPY)} = 149316 \times (\text{SIL})^{-3.36} \quad (7\text{-}23)$$

Figure 7.46 shows the results of correlations between SIL values and various aseismic methods on the fifth floor of the building.

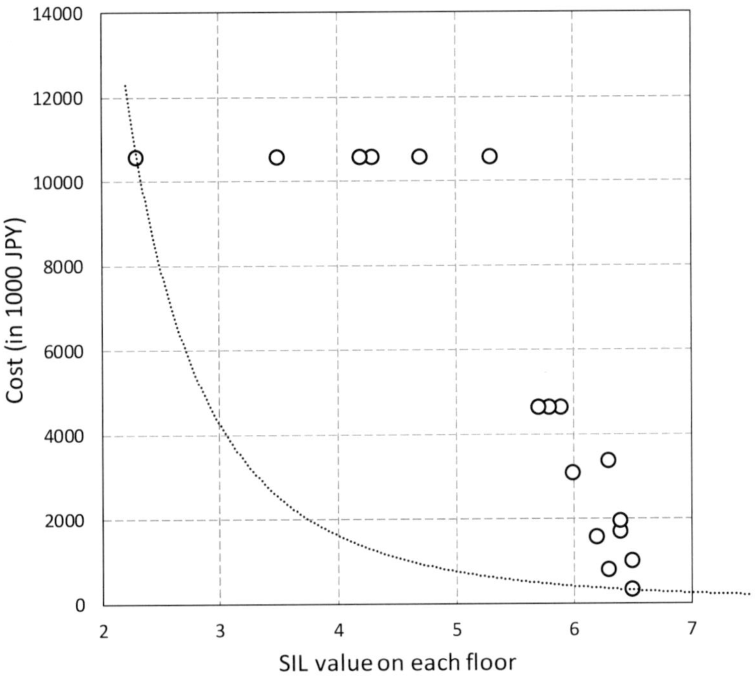

Figure 7.46 SIL and Cost of various aseismic methods on the fifth floor.

The trendline, which follows the power law of the optimal solution, is given as follows:

$$\text{Cost (in 1000 JPY)} = 179641 \times (\text{SIL})^{-3.402} \tag{7-24}$$

Figure 7.47 shows the results of correlations between SIL values and various aseismic methods on the roof of the building.

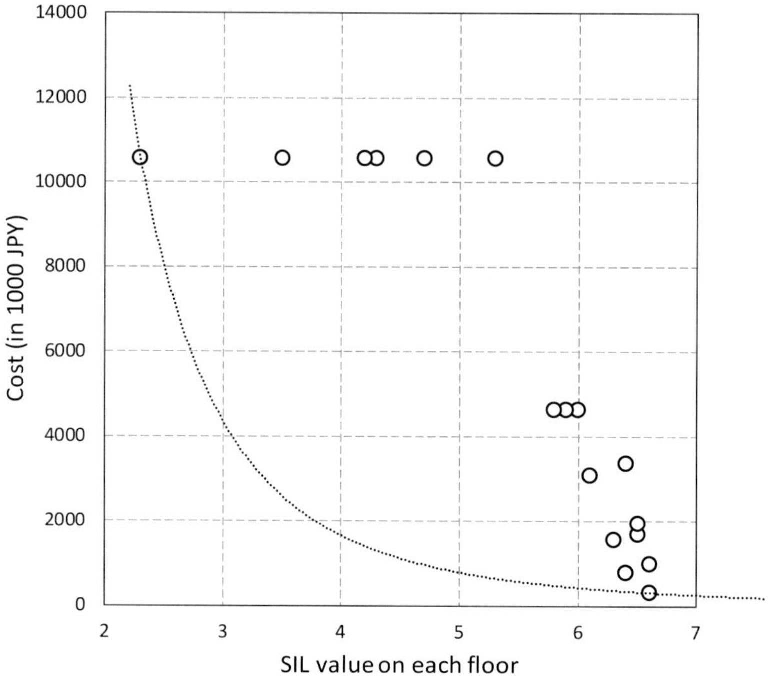

Figure 7.47 SIL and Cost of various aseismic methods on the roof.

The trendline, which follows the power law of the optimal solution, is given as follows:

$$\text{Cost (in 1000 JPY)} == 172418 \times (\text{SIL})^{-3.353} \qquad (7\text{-}25)$$

Figure 7.48 shows the power law of the optimal solutions of correlations between SIL values and various aseismic methods on all the floors and the roof of the building.

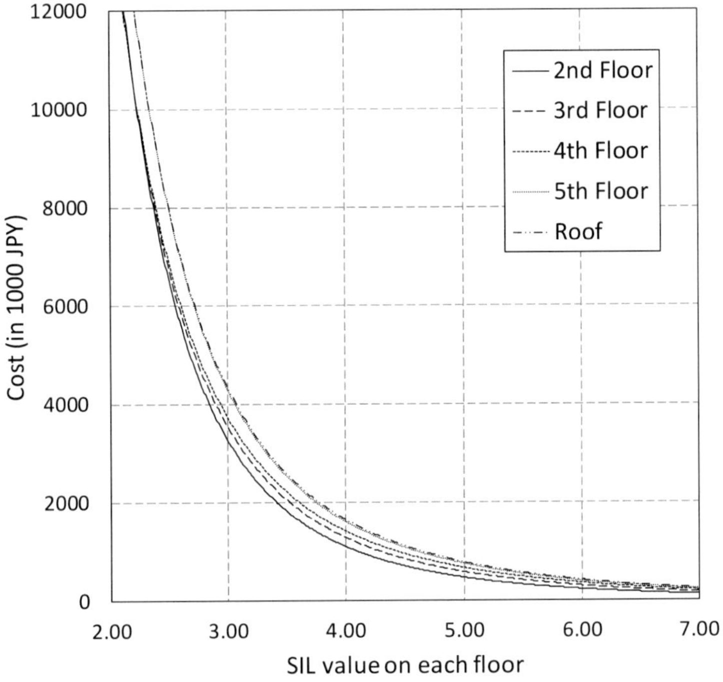

Figure 7.48 SIL and Cost of various aseismic methods of the building.

In Figure 7.48, the most optimum power law is found to be the second floor. Therefore Equation (7.21) is considered the representative solution for the prototype building used in this study.

From the curve, we observe that to gain a lower shaking level of SIL, we need to spend a higher cost for the advanced aseismic method. In other words, the lower the cost, the higher the value of SIL as a trade-off.

Finally, as a comment, without reducing the shaking, the building is already safe. Hence, reducing shaking during an earthquake and having a comfortable living environment are luxuries.

However, a reasonable cost of a base isolator for a light structure with a low number of stories can be achieved. Therefore, the role of SIL as an evaluation tool is a new finding in current practice that has been overlooked in planning, design, strengthening, and all aspects of building investigations.

References

[1] Kasai, K., Ooki, Y., Motoyui, S., Takeuchi, T., and Sato, E. E-Defense Tests on Full-Scale Steel Buildings: Part 1 – Experiments Using Dampers and Isolators, Structural Engineering Research Frontiers, 2007, ASCE, 12 p.

CHAPTER EIGHT

ARTIFICIAL INTELLIGENCE FOR SIL

Early warning systems of an earthquake

The early warning systems of the JMA and National Research Institute for Earth Science and Disaster Resilience (NIED) in Japan can disseminate earthquake information such as magnitude, location of the hypocenter, and Seismic Intensity Level (SIL) to a wide area surrounding the hypocenter immediately after a strong earthquake has occurred. Calculating the SIL uses data of between one to three seconds measurement periods of P-wave from at least four stations to predict the location, depth, and magnitude of an earthquake. However, the calculated SILs are done for all earthquake monitoring stations. Thus, the calculated SILs are the seismic intensities of the ground surface in all monitoring stations. These calculated SIL values obviously cannot represent the shaking intensities of every floor in a building above the ground, which will experience different or sometimes significantly higher levels of shaking.

To develop an early warning system of shaking intensities on all building floors that is faster and more robust during strong earthquakes very close to the earthquake's location, it is necessary to detect the SIL values rapidly. Therefore, it is necessary to develop a methodology to classify the shaking on each floor in a building during an earthquake occurrence.

The details of the calculation method to identify the shaking intensity of the ground during the earthquake are given in Chapter Five. To calculate the SIL values of each floor in a building, we need to conduct numerical analyses using a computer.

However, in the early warning system, to detect the shaking of a floor in a building, computer modeling and simulation has to be used to calculate the corresponding SIL appropriately. Even though computer technology is advancing rapidly, at the moment conducting computer modeling and simulation still requires several minutes, which is not fast enough to be used in real monitoring.

Therefore, a quick prediction of the SIL values in a building is necessary to use the early warning system for disaster mitigation purposes.

Artificial Neural Network

Inspired by the functionality and mechanism of biological neurons, an artificial Neural Network (NN) comprises a collection of simple processing elements (neurons or nodes) with multi-connectivity. As the input signal is received by each node, that is, the aggregate of the total information from other nodes or external stimuli (Figure 8.1). This signal is processed at the receiving node using a slow and imperfect activation or transfer function. The network is very efficient in performing large numbers of tasks.

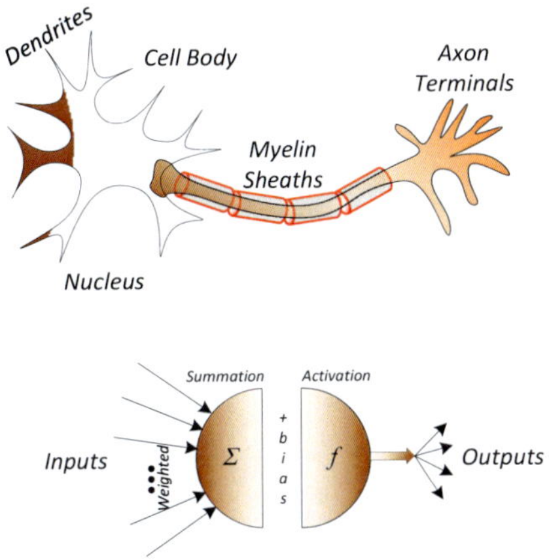

Figure 8.1 Biological and Artificial Neurons.

Some neurons may influence the environment by triggering actions. Learning is about finding influences that make the NN exhibit the desired behavior, such as recognizing a picture. Depending on the problem and how the neurons are connected, such behavior may require multiple connectivities among neurons. We live in the big-data period where all areas of science and technology generate massive amounts of data. This challenges us with extraordinary tasks regarding their interpretation and usage. For this reason, there is an urgent need for novel artificial intelligence and machine learning approaches to help us utilize these data. The Deep Neural Network (also known as Deep Learning, DL) method is a novel methodology consisting of a family of simple learning algorithms that can be used to learn complex problem predictions.

This chapter explores the application of the DL framework for rapidly predicting the extent of shaking on each floor in a building using the distance, depth, and magnitude of earthquake information (Figure 8.3).

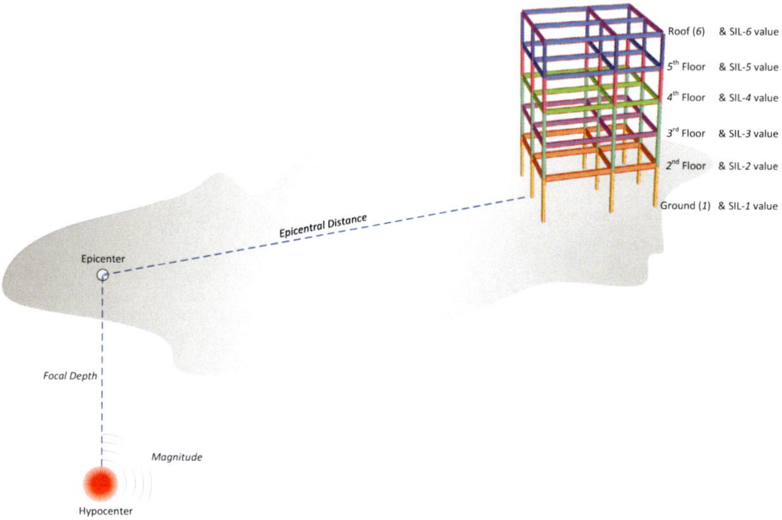

Figure 8.2 Earthquake information used in DL framework.

Building prototype

Big data on the level of shaking represented by SIL is not yet available. The calculation is still based on analyzing an existing building during strong earthquakes. Thus, the results of the predictions of SILs are a unique solution for each building. Hence, they cannot be valid for other buildings because they depend on the material type (steel, reinforced concrete, wood, etc.) and many factors that affect the behavior of the building, such as soil conditions, distance from the hypocenter, and others. A five-story steel frame building is the prototype used to narrow the problem. The building plan is 12m by 15m, and the total height is 18.1m. The building was designed by following the Japanese seismic codes. Figure 8.3 shows the real building prototype (steel structure) for generating big data of calculated SILs on each floor.

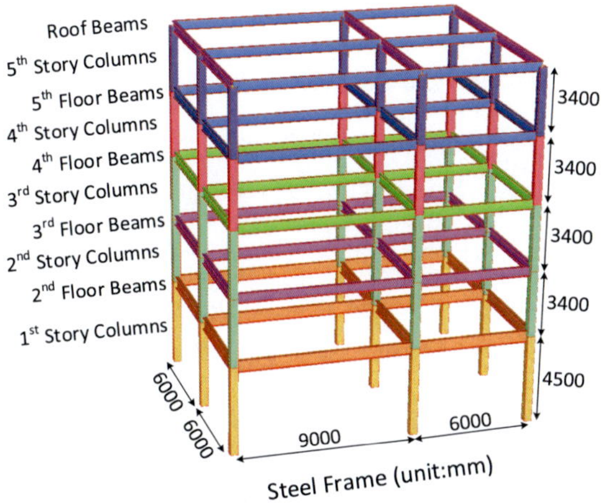

Figure 8.3 A five-Story Steel Frame Building.

The steel structure comprises wide-flange beams and square box column sections (Table 8.1). All the beams and columns have fully restrained connections.

Table 8.1 Cross sections of beams and columns.

Floor	Story	Beam	Column
Roof		W-500x200x9x16	
	5		□-400x400x12
5		W-500x200x9x19	
	4		□-400x400x16
4		W-500x200x9x22	
	3		□-400x400x19
3		W-500x250x9x22	
	2		□-400x400x19
2		W-600x250x12x22	
	1		□-400x400x22

The steel structure has been designed to meet the seismic design codes in Japan [1].

Test case

A test case is demonstrated herewith to explain how the real SIL values can be computed and what necessary information is required to predict a SIL value at an arbitrary location. The 2011 Great Tōhoku Earthquake was selected for this demonstration purpose. The earthquake occurred at 14:46 JST on 11 March 2011. The hypocenter of this earthquake was detected under the Pacific Ocean, 72 km east of the Oshika Peninsula. The ground surface of the hypocenter was located at a coordinate with a Latitude of 38.10 and a Longitude of 142.86, at a depth of 24 km. A magnitude of 9.0 (M_w) was reported [2].

In this test case, the nearest seismic station to the hypocenter, MYGH03 station, was selected. MYGH03 station is located at a coordinate with a Latitude of 38.92 and a Longitude of 141.64. The station is at 80 m height above sea level. The direct surface distance from the hypocenter is about 163.5 km. The strong motion seismic data was obtained from the K-NET and KiK-net seismic network archives, managed by the National Research Institute for Earth Science and Disaster Resilience (NIED), Japan.

The strong motion data recorded at the station consisted of N-S, E-W, and U-D directions of seismic waves. The vector of resultant acceleration waves was used as the computer's ground motion input for earthquake analysis (Figure 8.3). The accelerometer used to record the seismic waves was operated at 100 Hz (0.01 second). In the earthquake analysis (Figure 8.4), the time increment used one-fifth of the data increment time, thus 0.002 seconds for 300 seconds (150,000 increments) as the total analysis time.

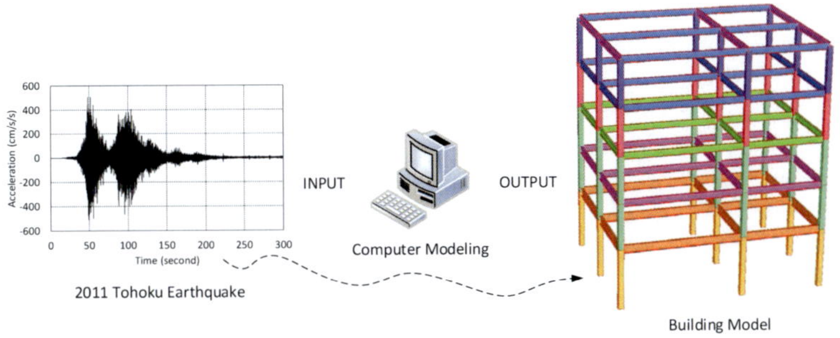

Figure 8.4 Earthquake Analysis Simulation.

The results of response accelerations on each floor of the building model from the earthquake analysis simulation are summarized in Table 8.2 and illustrated in Figure 8.5.

Table 8.2 Results of m-SIL calculations of the existing building.

Floor	m-SIL	T (sec) [FFT]	$A_{0.3}$(gals)	Natural Frequencies of Building Prototype
Roof	6.6	0.2606	1393.8	Mode 1: T = 0.2621 sec Mode 2: T = 0.2173 sec Mode 3: T = 0.1870 sec Mode 4: T = 0.0942 sec Etc.
5	6.4	0.2606	1107.1	
4	6.2	0.2606	879.4	
3	6.1	0.2606	783.8	
2	5.8	0.2606	554.9	
Ground	5.2	0.6836	164.4	

It can be seen in Table 8.2 that the $A_{0.3}$ accelerations and calculated SIL values are increasing in the height direction.

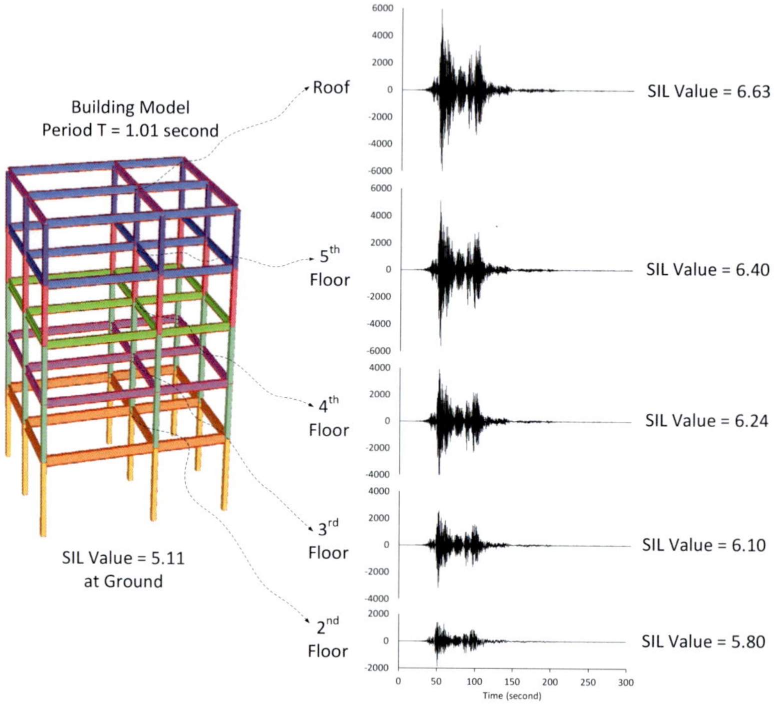

Figure 8.5 Computed shaking intensity on each floor using a computer model.

From the calculated m-SIL values, the dominant period of this building (1.01 seconds) can be computed by following the procedure described in Chapter Five. By plotting the values of the dominant period and $A_{0.3}$ accelerations in the SIL graphs, we can see that the SIL at the ground surface is much less than on the roof.

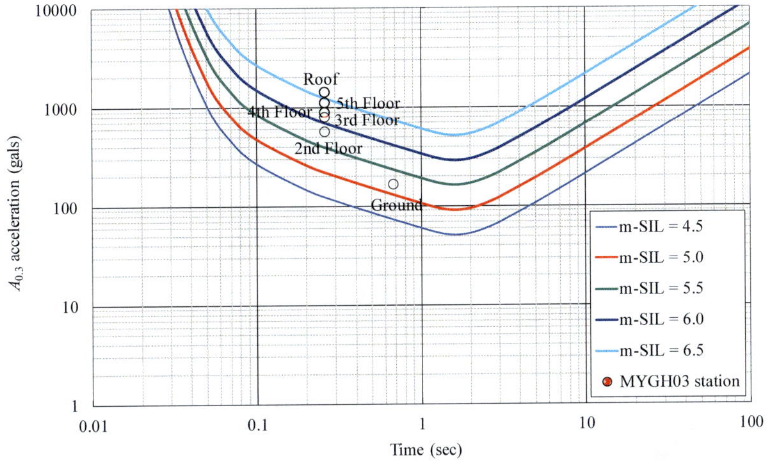

Figure 8.6 m-SIL of a building prototype located at MYGH03 Station.

As demonstrated above, we can determine the m-SIL values on each floor in a building by using the seismic wave recorded in one seismic station. We can observe that the SIL value being broadcasted as an early warning system in Japan is actually the SIL value of the ground surface. The value does not represent the shaking of each floor in the building; that is obviously higher than the SIL on the ground surface.

The building example used in the above example will be used as a prototype to generate the calculated m-SIL values from many seismic stations.

Earthquake data source

We compiled strong motion data from K-NET and KiK-net seismic network archives managed by NIED, Japan. The strong motion data were collected based on the value of SIL from 5 to 5.8. After searching the strong motion data archives, 353 seismic stations with resultant vectorized 552 accelerations (from 552x3 accelerations records in three directions: N-S, E-W, U-D) were recorded from 23 May 1996 to 16 March 2022.

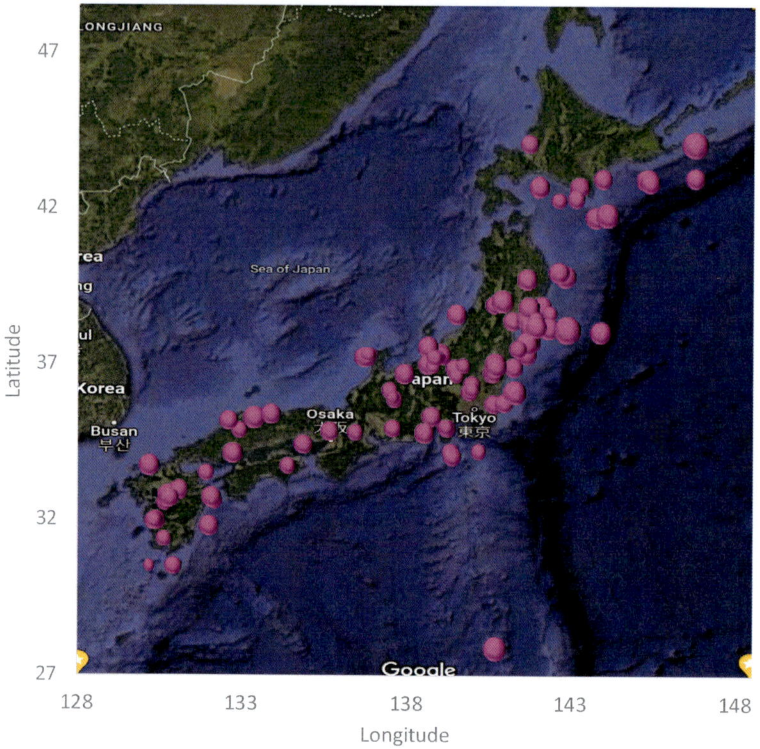

Figure 8.7 The hypocenters of selected 552 earthquakes, 1996–2022, plotted on the satellite view of Google Maps.

Figure 8.7 shows the 552 hypocenters of earthquakes recorded in 353 seismic stations in Japan from 23rd May 1996 to 16th March 2022. Figure 8.8 shows the locations of the seismic stations from where the strong motion data were taken.

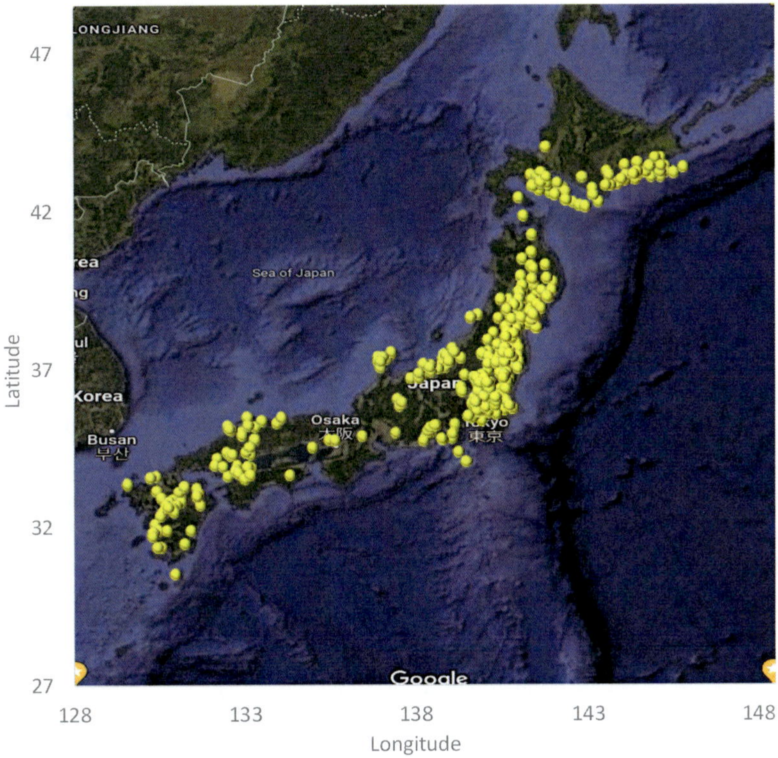

Figure 8.8 353 Seismic Stations recorded SILs from 5 to 5.8 (1996–2022) plotted on the satellite view of Google Maps.

Earthquake and SIL dataset

A dataset in DL is a collection of information that can be treated by a computer as a single unit for diagnostic and prediction purposes. Then, the information should be uniform and understandable for a machine that does not see the information in the same way humans do. Therefore, after composing the information, it is important to preprocess it by placing it in an appropriate arrangement and annotating it by adding meaningful tags readable by a computer.

The resultant vectorized seismic waves from 353 seismic stations comprised earthquakes with SIL values between 5 to 5.8. A total of 552 seismic waves will be used as the input of earthquake loadings to the building prototype described in the Test Case. As shown in the Test Case above, for one input seismic wave at the ground, there will be five response accelerations as the results of the earthquake simulation.

Altogether with the calculated SIL values and floor data, the dataset will consist of information (Figure 8.9) on:

1. Distance between the location and the epicenter
2. Depth of the hypocenter
3. Magnitude of the earthquake
4. Floor identification
5. SIL value of a floor

As, for one earthquake, after conducting the earthquake simulation of the building prototype, the ground surface and floors in the building will have six different SIL values, resulting in a dataset of 552× 6=3,312 items.

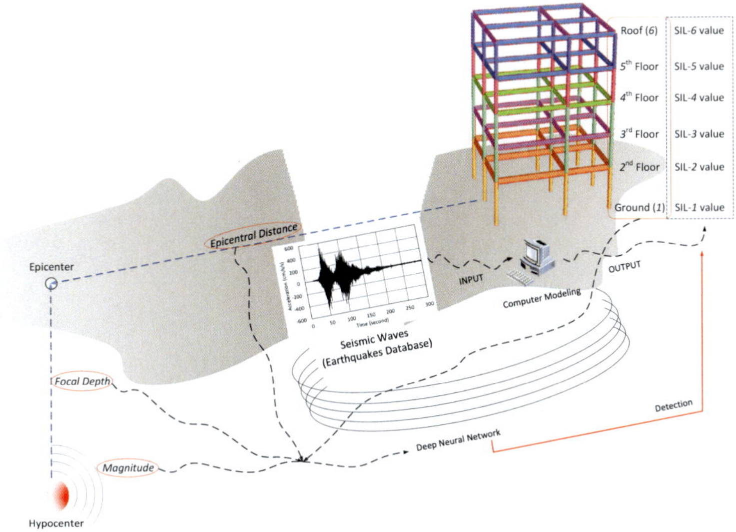

Figure 8.9 Computer Simulation of a building during an earthquake.

Deep Neural Network (DNN)

DL is one machine learning technique that learns features directly from the dataset. When the amount of the dataset increases, DL will perform with better accuracy.

Most of the applications of DNN in this real world are found in image identification and speech recognition analyses because they have massive amounts of information that can be considered big-data mining. On the contrary, the application of DNN in the exact field is not imminent because of the solution in nature. Hence there is no need for prediction.

In earthquake engineering, the characteristics of an earthquake, such as magnitude, hypocenter, and others, become the subject of prediction because many uncertainties cannot be solved by the exactness of science and technology.

Dataset structure

The dataset of SIL is inspired by the breakthrough results in image analysis using DNN. The dataset of an image is constructed following the image resolution; thus, a square matrix is usually assumed necessary to create a dataset.

The structure of Earthquake Building and SIL datasets is shown in Figure 8.10, where the matrix size is arranged to be the size of 3x3 and 1x1 size for the SIL values. This SIL value will be used to verify the predictions. The floor identification is given three diagonal slots (No.4) to emphasize that the SIL value belongs to the floor dataset.

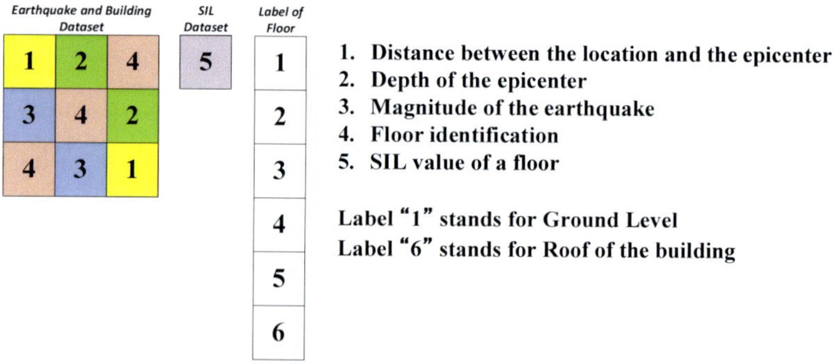

Figure 8.10 The structure of the SIL dataset.

It should be noted that there is no SIL solution for the pre-occurred earthquake because that is not in the dataset. The prediction of SIL at a particular location due to an unknown earthquake hypocenter can only be verified when the earthquake has occurred.

DNN model

A DNN model must be designed to predict floor labels and SIL values. Figure 8.11 shows the DNN model that consists of the following:

- A convolution-batch normalization-ReLU layers with 16 2-by-2 filters.
- Two convolution-batch normalization-ReLU layers with 32 2-by-2 filters.
- A convolution-batch normalization-ReLU layers with 64 2-by-2 filters.
- A skip connection from the mainstream contains convolution-batch normalization-ReLU layers with 32 1-by-1 filters.
- A convolution-batch normalization-ReLU layers with 64 2-by-2 filters.
- Merge the skip connection using the addition operator.
- For the classification output of Earthquake Building prediction, a branch with a fully connected operation of size 6 (the number of floor classes) and a softmax operation.
- For the regression output of SIL value, a branch with a fully connected operation of size 1.

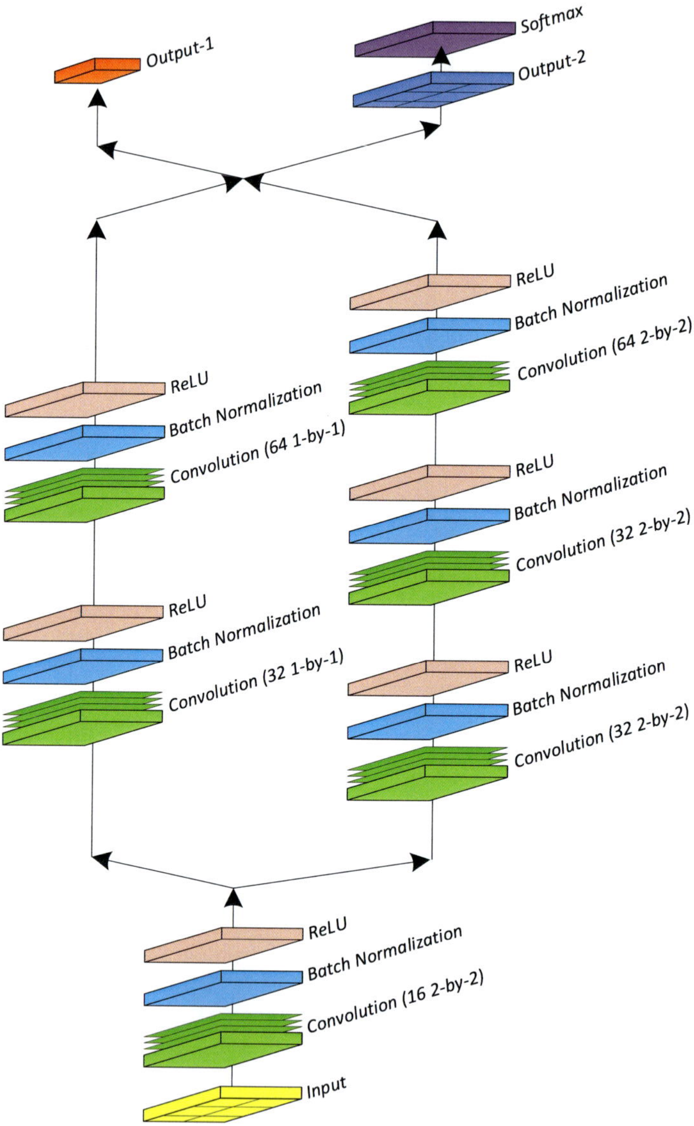

Figure 8.11 DNN model of hidden layers.

Convolution in deep neural networks

Convolution neural networks are widely used tools for deep learning. Convolution neural networks are inspired by the biological structure of a cortex, which contains arrangements of simple and complex cells of a layer of tissues. These cells are active based on complex cell subregions. These subregions are called receptive fields. The subregions are constructed by sliding a region like a window in the entire region (Figure 8.12). Inspired by the findings, the neurons in a convolutional layer connect to the subregions of the layers before that layer instead of being fully connected as in other types of neural networks. The neurons are unresponsive to the areas outside of these subregions in the region.

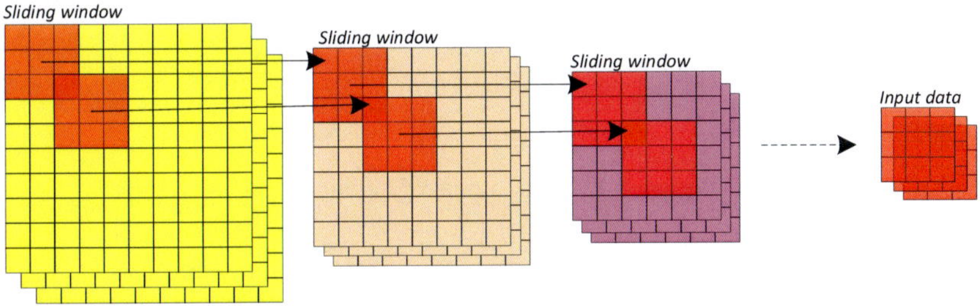

Figure 8.12 Convolution of a region in a dataset.

Training data

From the 3,312 recorded and calculated data items, 3,000 datasets were used to train the DNN model. The model was trained for 30 epochs using a mini-batch size of 64. For each epoch, the data was shuffled and looped over the mini-batches region. At the end of each iteration, within each mini-batch:

- The mode-loss and gradients of the data were evaluated

- The network parameters were updated

The results of loss during the training are shown in Figure 8.13.

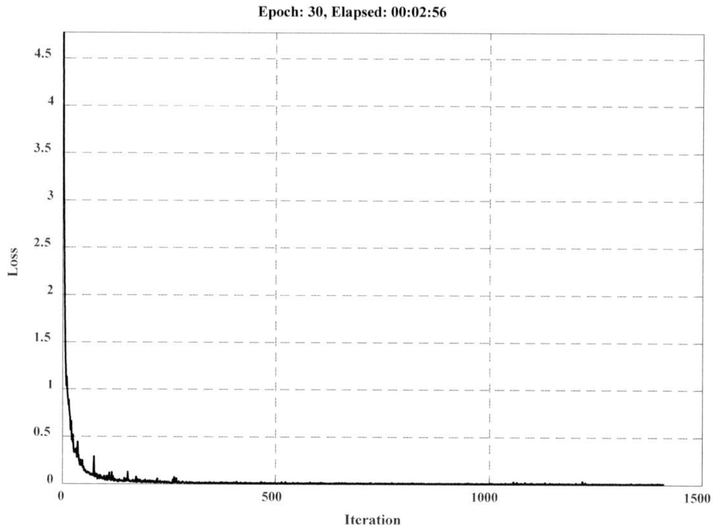

Figure 8.13 DNN model of hidden layers.

It can be seen from Figure 8.12 that the convergence of the DNN model was very fast to yield nearly zero loss with less than 500 iterations.

Validating data

Data validation was conducted to test the classification accuracy of the remaining 312 datasets in predicting the floor label and SIL values using the 3,000 trained datasets. The evaluation of floor labels accuracy yielded nearly 100% accuracy. In contrast, the regression evaluation of the SIL values predicted in terms of Root Mean Square Error (RMSE) gave a value of 0.4445.

It is impossible to obtain a nearly zero RMSE value because there is no equal solution between the validation and training data. The results of discrepancies in the validating process of each floor in a building are shown in Figure 8.14.

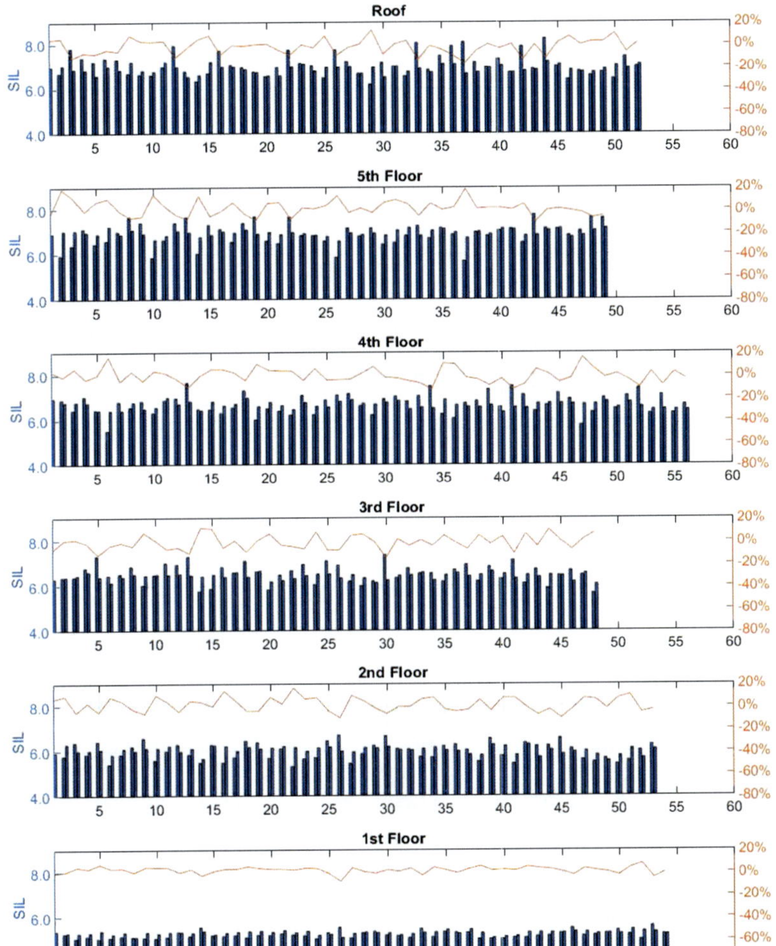

Figure 8.14 Statistic SIL value predictions of building prototype.

The results of SIL values predictions of the validation dataset are summarized in Table 8.3.

Table 8.3 Predictions of SIL using validation dataset (in one random run).

Floor	Min SIL (%)	Max SIL (%)
Roof	6.3 (-0.2)	7.2 (+0.1)
5	6.3 (-12.8)	7.2 (+17.2)
4	6.3 (-10.2)	7.0 (+21.1)
3	5.8 (-16.1)	6.6 (+11.2)
2	5.6 (-14.0)	6.4 (+11.9)
Ground	5.1 (-4.0)	5.5 (+8.7)

From Table 8.3, we can observe that the SIL values increase with the height of the building. The predictions at the ground level and roof of the building show a good relationship between the trained and validating datasets (less than 10% error). The predictions for the other floors show a higher error percentage (about 20%).

Prediction of SIL

The summary of 3,312 trained and validated datasets are shown in Table 8.4. The strong motion data are collected based on the value of SIL from 5 to 5.8. The distance represents the direct surface locations between the hypocenter and seismic stations. The depth represents the vertical distance between the depths of hypocenters to the Sea Water Level height of seismic stations.

Table 8.4 Summary of 3,312 dataset ranges in DNN of a SIL model.

Type	Min	Max
Distance	1.97 km	519.3 km
Depth	4.05 km	108.4 km
Magnitude	3.9	9.0

To see how good the DNN model is, twenty random datasets consisting of distance, depth, and magnitude of earthquakes were considered. The datasets were constructed within the minimum and maximum values, as shown in Table 8.4.

Figures 8.15 to 8.20 show the results of SIL predictions of the building prototype. The figures plot the magnitude (Mw) of twenty random earthquakes compared to the SIL values.

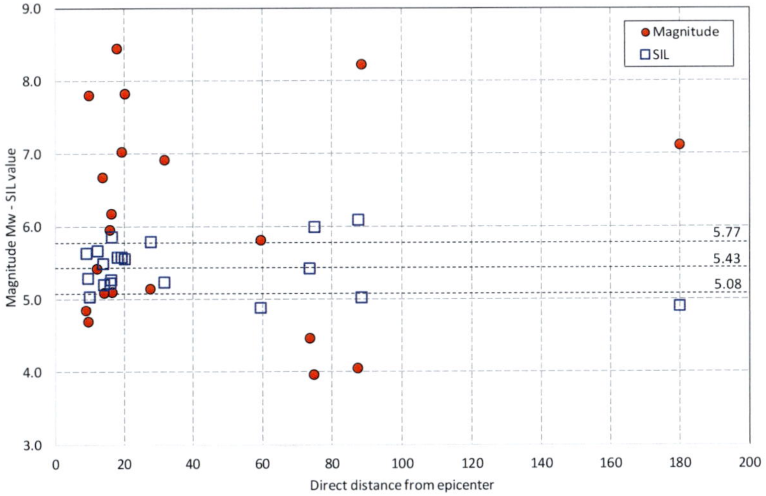

Figure 8.15 Predicted SILs at the ground surface of the building prototype.

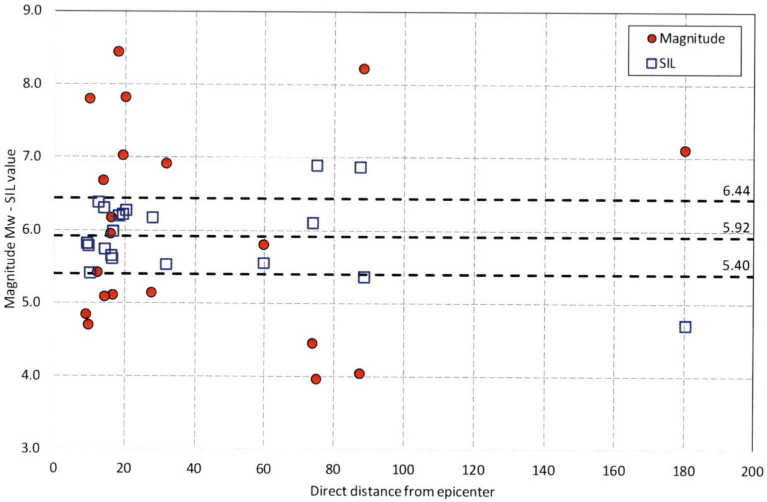

Figure 8.16 Predicted SILs on the second floor of the building prototype.

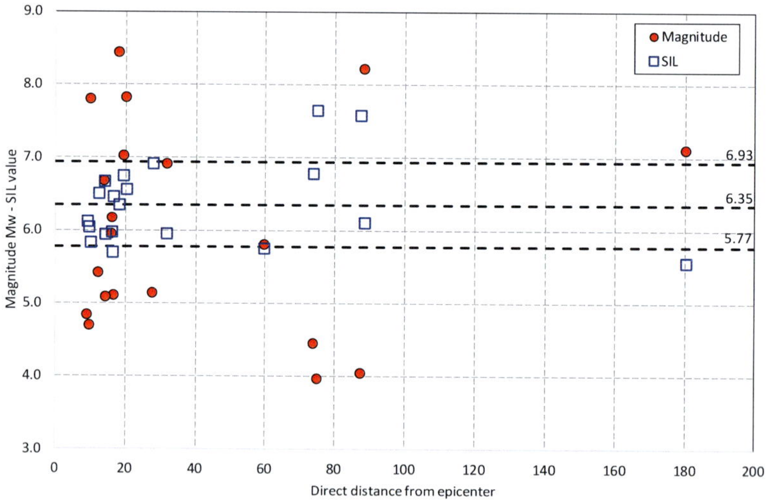

Figure 8.17 Predicted SILs on the third floor of the building prototype.

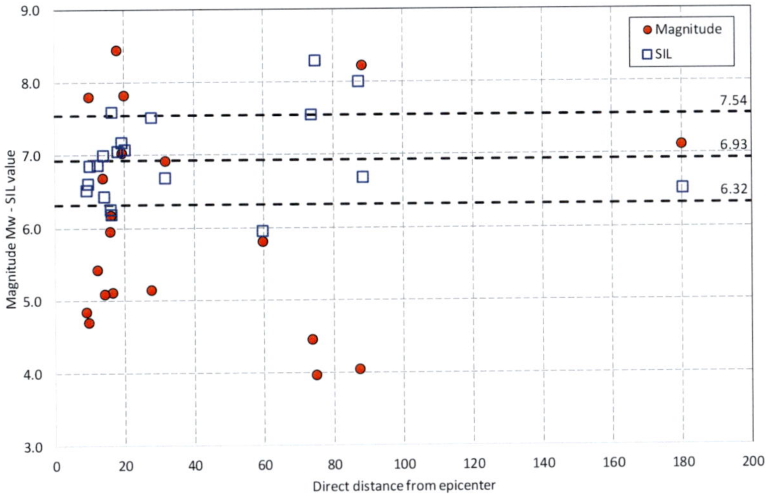

Figure 8.18 Predicted SILs on the fourth floor of the building prototype.

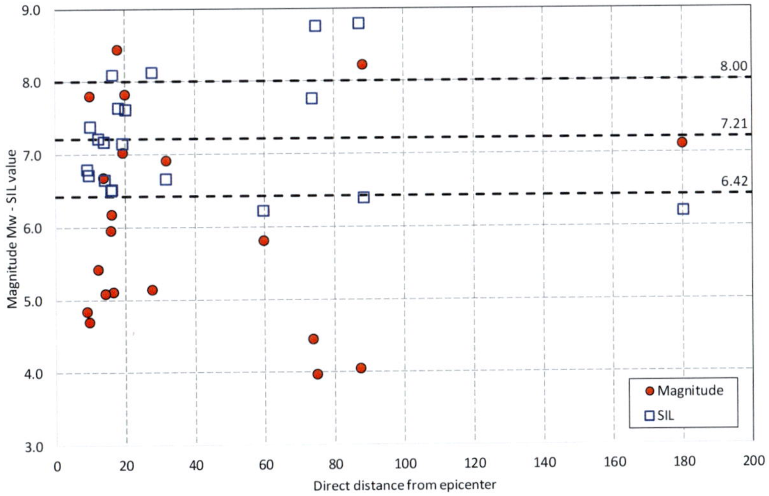

Figure 8.19 Predicted SILs on the fifth floor of the building prototype.

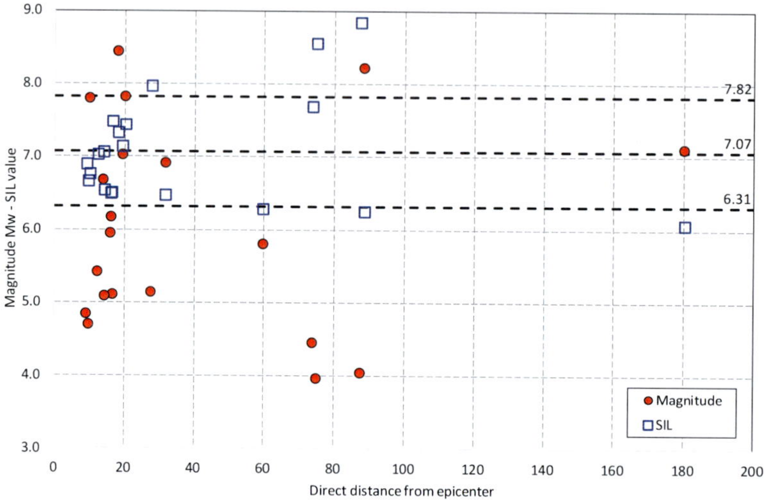

Figure 8.20 Predicted SILs on the roof of the building prototype.

In the figures, the average values of SIL on each floor of the building prototype are depicted in the middle of three horizontal dashed lines. The upper and lower horizontal dashed lines show the range of twenty sample standard deviations.

The predicted SILs with the $A_{0.3}$ acceleration (Figure 8.5) on each floor of the building prototype are summarized in Table 8.5.

Table 8.5 Summary of SIL prediction using twenty random datasets.

Floor	Predicted SIL (average)	Standard Deviation	T (sec) [Mode-1]	$A_{0.3}$(gals)
Roof	7.07	0.75	0.2606	2,394.4
5	7.21	0.79	0.2606	2,813.2
4	6.93	0.61	0.2606	2,038.0
3	6.35	0.58	0.2606	1,045.2
2	5.92	0.52	0.2606	637.1
Ground	5.43	0.35	0.6836	214.2

The calculated $A_{0.3}$ accelerations on each floor of the building prototype are plotted in the SIL graphs, as shown in Figure 8.21. The dominant time of the building prototype is assumed to be similar to the existing building behaviors, which are close to the natural frequency of the building in Mode 1 (Table 8.2). The figure shows that the higher floors show larger SIL values, although the fifth floor is slightly larger than the roof.

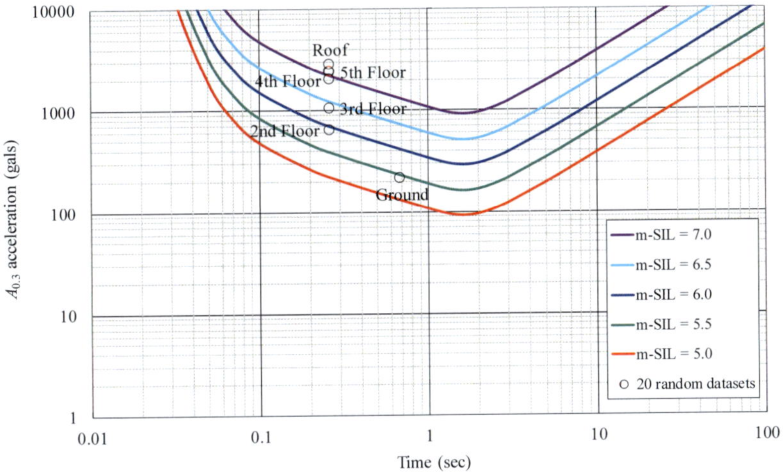

Figure 8.21 Predicted m-SILs statistics of the building at twenty random locations.

When the predicted SILs of the building prototype and magnitudes of the random earthquakes are plotted together in Figure 8.22, we can see no correlation between the magnitudes and SILs, because the results are scattered, not in a diagonal manner. There are cases where Mw > SIL (typical SIL at the ground surface, see also Figure 4.2), but also there are cases where SIL > Mw values even in the lower range of magnitude (4–5). These results imply that even in a medium-scale earthquake of 4–5, the shaking of a building, like the building prototype adopted in this study, can be amplified significantly (SIL>7.0). Therefore, a high risk of casualties requires an early warning system and disaster countermeasures as well as thoughtful considerations in the early stages of the planning and design of a building.

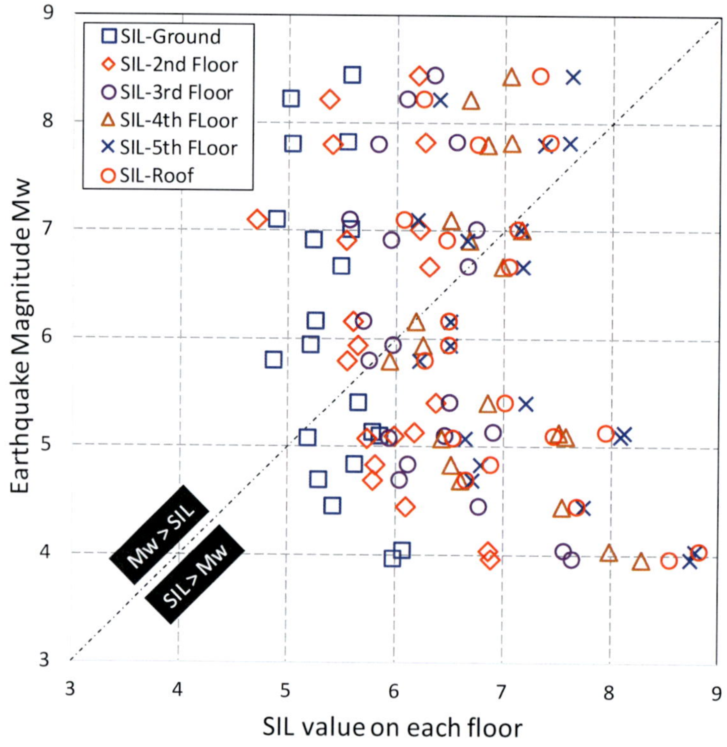

Figure 8.22 Correlation between predicted SIL and Mw values.

References

[1] Kasai, K., Ooki, Y., Motoyui, S., Takeuchi, T., and Sato, E. E-Defense Tests on Full-Scale Steel Buildings: Part 1 – Experiments Using Dampers and Isolators, Structural Engineering Research Frontiers, 2007, ASCE, 12 p.

[2] NIED K-NET, KiK-net, National Research Institute for Earth Science and Disaster Resilience, doi:10.17598/NIED.0004.

CHAPTER NINE

SIL DETERMINATION USING EEG & AI

Comfort and discomfort

Diversity of backgrounds, such as race, gender, and age, of a human being means they may have different feelings, even in the same situations and conditions. However, most evaluation methods in comfort are based on a fixed standard without considering human background diversities. In this chapter, we emphasize the preference for comfort. We discuss how the dipole imaging of human brain waves can be used to evaluate comfort based on SIL values. Human beings spend most of their lives improving learning efficiency and work productivity, which leads to a comfortable and healthy environment [1].

In response to environmental stimulations, humans express their emotions differently. Therefore, it is not easy to categorize the model of human emotions. Until 1980, the model of human emotions was defined linearly between two extreme conditions: happiness and sadness; relaxation and panic; excitement and boredom, etc. In 1980, Russell [2] proposed a two-dimensional model of emotions separated into four quadrants separated by two axes of arousals (low-high) in the horizontal direction and valences (negative-positive) in the vertical direction, as shown in Figure 9.1. According to this model, valence denotes the quality of an emotion ranging from unpleasant to pleasant. While arousal denotes the quantitative activation level of emotion ranging from calm to excitement.

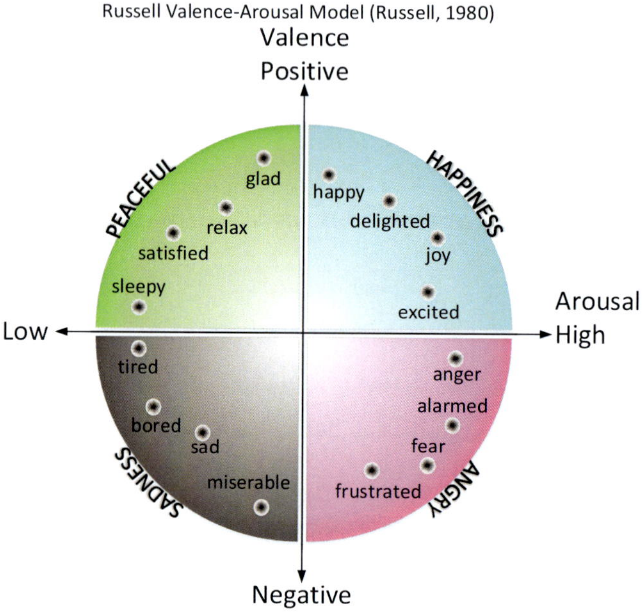

Figure 9.1 Two-Dimensional Model of Emotions.

To understand the emotions of human beings, the most important basic psychological characteristics to consider are comfort and discomfort.

Shaking and emotion

Many works of literature have used stimuli such as movies, music, and text to elicit the emotions of a human being. Among these stimuli, movies have the looked-for properties of being readily standardized and involving no dishonesty. Dynamic visual and audio stimuli make movies seem to be one of the most effective ways to elicit the emotions of a human being. However, no literature relates the relationship between the shaking due to earthquakes and the emotions of humans that feel the shaking.

The development of using shaking as an emotional elicitor has paralleled the broader growth of emotion science. It is interesting to study the differentiation of emotional states relating to shaking due to an earthquake. The emotions of human beings related to the shaking will benefit seismic design practices. In current seismic design practices, only the "strength" of a building is the target to be accomplished. Comfort is not given consideration in the design practices since the design codes are put into the enforcement of building and construction laws. Hence, there should be a comprehensive method of evaluating the extent of shaking of a building during the occurrence of an earthquake. The extent of shaking should be added to the current seismic design practices to allow designers to minimize the shaking even though the building is strong against earthquakes.

EEG and emotion

Various types of devices can measure emotion in the human brain, such as Magnetic Resonance Imaging (MRI), magnetoencephalography (MEG), functional Near-Infrared Spectroscopy (fNIRS), or an electroencephalogram (EEG). MRI and MEG have an advantage in high spatial resolution, but they are slow in resolving ability. On the other hand, the drawbacks of fNIRS and EEG are low spatial resolution and the difficulty in measuring deep brain activity. However, the devices are portable compared with MRI and MEG. They can measure human brain activities simply using a headgear-like device. Deep brain activity is not important for measuring the emotion of a human being during a shaking stimulus. For this reason, it is recommended to evaluate comfort considering individual variances using EEG.

Because the emotional system of the human brain is a basis for emotion-related processes, EEG can reveal important information on brain activities. There are many studies on correlations between EEG activities and the emotional system of the human brain. Russell [2] reported that greater values of left frontal EEG activities are associated with positive emotions. In comparison, greater right frontal EEG activity values are associated with negative emotions (Figure 9.1).

Measuring EEG activity

The human brain is a network of immeasurable nerve cells, which are thought to have more than 14 billion cells. It is believed to be the uppermost center that controls human thoughts and behaviors and people's emotions and autonomic roles. The neurons communicate through weak electricity via dendrites emerging from neurons. This phenomenon occurs in the cerebral cortex following the pyramidal orientation of cells, and the electrical activities of neurons are superimposed on each other and transmitted to the scalp. The EEG is a system of the measurement of transmitted electrical activities (Figure 9.2).

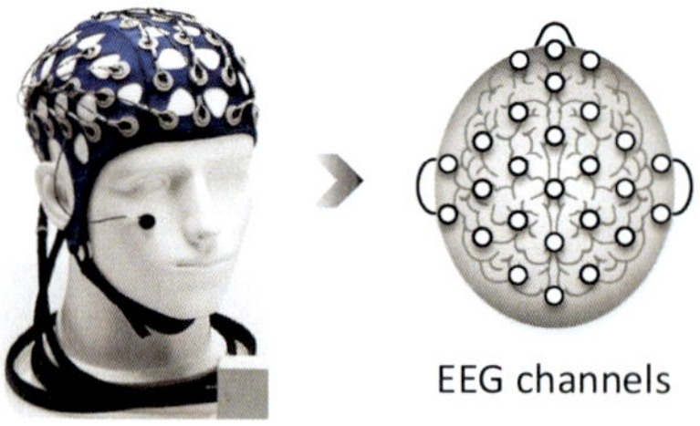

Figure 9.2 EEG measurement channels.

The electroencephalogram effectively explains brain function in a near-natural situation because the measurement situation is not restricted and can be easily measured without interruptions. However, due to the limited number of electrodes and the low conductivity of the human skull, the spatial resolution of EEG is low. Hence, it is not easy to identify the electrical activities in the human brain directly from the potential distribution in the scalp region. The equivalent dipole signal strength distribution of the scalp potential and the signal source can be quantified to identify the electrical activities in the human brain. The signal source generated from the brain can be equivalently expressed by the distribution of multiple dipole signal intensities on the simulated surface in the brain without being limited in the number and direction (Figure 9.3).

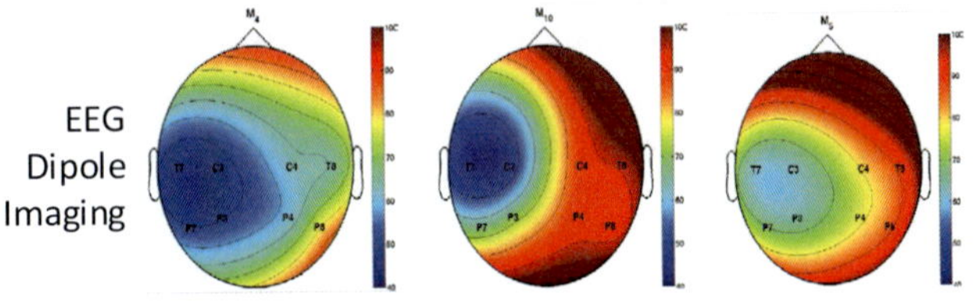

Figure 9.3 Illustrative dipole imaging of EEG activities.

Proposed experiment using EEG

The original SIL values were determined based on tree movements and houses trembling during strong earthquakes. Four categories of SIL levels have been used by JMA for a long time, since 1884. At that time, there were only four seismic intensity shaking values, "Small," "Weak," "Strong," and "Severe." After 1949, after experiencing large earthquakes, the need to refine the SIL levels for early warning systems yielded seven seismic intensities. Since 1996, the present SIL levels have been further refined to ten levels, and the method used to calculate SIL was augmented. Although the formula to calculate SIL was established and used widely in the country, it was an empirical equation based on the 0.3 seconds from the maximum response of the ground surface. However, the refinements to the seismic intensity levels so far were based on human perceptions qualitatively during a strong earthquake. Therefore, there were no correlations between the SIL and its formula, which was not based on technical sciences as the decision-making background.

Therefore, it is necessary to determine the levels of SILs based on scientific judgments of human perception of shaking due to strong earthquakes. Also, the formula should be correlated with human perceptions in determining the corresponding SIL values. To achieve this, the author proposed the following experiment to obtain data from the shaking table experiment.

An experiment to observe electrical signals from the human brain during the shaking stimulus, as shown in Figure 9.4, is proposed.

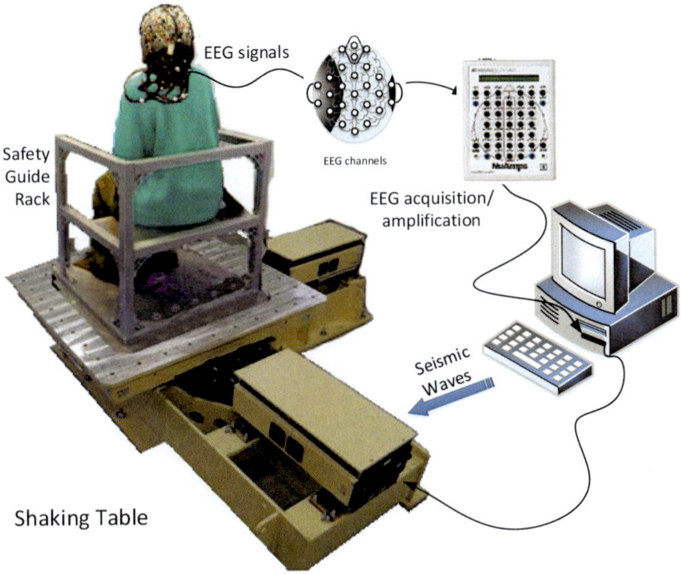

Figure 9.4 Shaking table test using EEG.

The experiment results can be represented in the dipole imaging, as illustrated in Figure 9.5. Figure 9.5 shows the parietal hemisphere after normalizing the dipole signal intensities. The upper part shows the frontal human scalp images. Observing Figure 9.5, the frequencies of the signal measured by EEG can be classified into five main frequency windows. The frequency windows are Delta (0–4 Hz, Deep Sleep), Theta (4–8 Hz, Light Sleep), Alpha (8–12 Hz, Relaxed), Beta (12-30 Hz, Active/Alert), and Gamma (30 Hz, Hyperactive/Alert) in ranges with the emotional conditions of a human being.

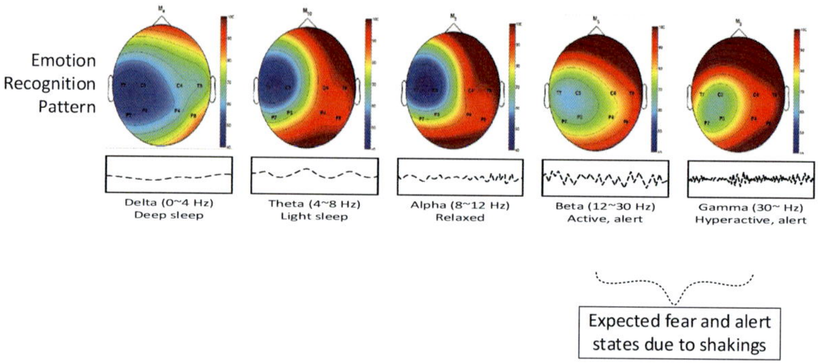

Figure 9.5 Example of experiment results using EEG.

When the human brain emits a signal frequency of about 12 Hz or more, the brain's emotion indicates it is in an active or alert condition. Although the border distinction of the functional conditions may differ between humans with age, gender, or other influencing characteristic factors, the questionnaires and visual observations of the reactions of humans during the experiments can be bound together to detect the moment when they feel the emotions of fear and alert states due to shaking.

Artificial intelligence for SIL using EEG

The dataset for identifying SIL values is composed of time history records of signals from each channel of the EEG device. The time history records of signals from EEG channels will be used as the inputs for DNN of emotion recognition correlated with SIL values during the shaking table experiments (Figure 9.6).

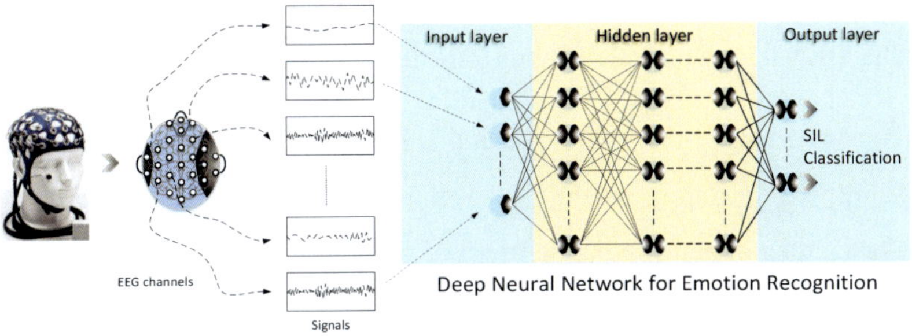

Figure 9.6 Application of DNN for SIL Classification.

Because we know the input of the seismic wave to the shaking table, we can correlate the responses of human beings in the experiments with the calculated SIL values. Hence, the verification of the validity and reliability of the formula to compute SIL might be calibrated for better predictions.

Future directions

Identifying SIL values without dynamic structural analysis, other than the ground surface, is important in the early warning system during earthquake occurrences. The SIL values for a building with several floors will differ depending on the height. However, detecting the SIL values on each building floor is cumbersome. Thus, it is not an efficient detection tool for disaster mitigation purposes.

Rapid developments in wearable and low-cost devices to measure EEG data have led to the development and leveraging of innovative devices to make measurements normally confined to the disaster mitigation management system. Aside from the measurement of SIL values on the ground surface, which already provides well-accepted endpoints for the early warning system, the technologies to detect the SIL values on each floor in a building are emerging and are the subject of continued research providing validation of the SIL values which are calibrated by using the experimental data.

It is acknowledged that more research is needed to better understand the reliability and validity of the SIL values and to generate the required understanding of new predictions derived from human brain data.

References

[1] Vernon, H.M. (1919). The influence of hours of work and of ventilation on output in tinplate manufacture, H.M. Stationery Off. London.

[2] Russell, J.A. (1980). A circumplex model of affect, Journal of Personality and Social Psychology, 39(6), 1161–1178.

INDEX

F